Lab Manual
to accompany:

The Science of Animal Agriculture

Frank Flanders
with Ray Herren

For information, address Delmar Publishers Inc.
3 Columbia Circle, Box 15015
Albany, New York 12212

Printed in the United States of America
published simultaneously in Canada
by Nelson Canada,
a division of The Thomson Corporation

2 3 4 5 6 7 8 9 10 XXX 00 99 98 97 96 95 94

Library of Congress Catalog Card Number: 92-21483

ISBN: 0-8273-5926-8

Lab Manual
to accompany:

The Science of Animal Agriculture

Frank Flanders
with Ray Herren

Delmar
Publishers Inc.™

I(T)P™

Table of Contents

Chapter 10: Animal Behavior

Chapter 11: Animal Genetics

Chapter 12: The Scientific Selection of Agricultural Animals

Chapter 13: The Reproduction Process

Chapter 14: Animal Growth and Development

Chapter 15: Animal Nutrition

Chapter 16: Meat Science

Chapter 17: Parasites of Agricultural Animals

Chapter 18: Animal Diseases

Preface

Agriculture is the application of science to the production of food and fiber. With the explosion of biological knowledge in the twentieth century the impact of the application of scientific research in agriculture has been tremendous. For students of agriculture, agricultural producers, and researchers this has meant a race to keep up-to-date with the applications of science in agriculture.

This laboratory manual accompanies the text *The Science of Animal Agriculture*. The manual is intended to help students appreciate the interrelationship between science and agriculture. These exercises are only a small sample of the extensive applications of biological principles to agricultural production. A lab manual on animal agriculture cannot be comprehensive — the field of study is too broad. The lab exercises focus on specific aspects of animal agriculture, but they are intended to be a representative sample of the field of study.

There are hundreds of experiments in animal agriculture that can be used with the text. The text and teacher's manual contain additional ideas for laboratory exercises. The teacher is encouraged to supplement the activities in this manual with those from the text and teacher's manual or those of his/her own creation. The teacher may want to modify the exercises in this manual to make them more relevant to the local community needs and circumstances.

This lab book contains thirty-six exercises directly related to the eighteen chapters in *The Science of Animal Agriculture*. Each exercise emphasizes and expands on scientific principles covered in the text. A key to this lab manual is that it is not a workbook. The exercises are, as far as possible, designed to be hands-on. In agricultural education the motto is "learning by doing." Students learn best from their own experience of doing practical science and work projects.

One objective of the lab manual, as with the text, is to give guidance in assisting students to deal on a personal level with the contemporary ethical issues in production agriculture, environmental issues, research, and agricultural processing. Efforts were made to present the issues from all standpoints in a balanced approach.

Follow-up study is encouraged. These exercises can be a catalyst for larger-scale student and class agriscience projects. Many awards programs in agriculture and science reward outstanding work, and developing scientific curiosity is the key to exciting careers in agriscience.

The exercises can easily be used for individual and small-group work in the agriscience laboratory with minimal assistance and supervision from the teacher. The teacher's main roles will be in preparation, providing preliminary information, and safety precautions. The exercises are designed for students to take responsibility for what they learn. There are many opportunities for students to answer their own questions through discovery learning and problem solving.

Acknowledgments

The authors gratefully acknowledge the following for their assistance in preparing this laboratory manual.

Dr. Dale Carpentier — for contributing excellent ideas and outlines for agriscience laboratory exercises.

Theresa Probst — for her assistance in writing and editing and suggestions for science-based activities.

Wayne Probst — for writing the laboratory activities on aquaculture, and for his excellent ideas for showing the relationship between science principles and agricultural applications.

NAME:______________________ DATE:____________ CLASS:____________

Chapter 1: Animal Agriculture as Science

Laboratory Exercise 1.1 The Scientific Method

BACKGROUND

Most of the comforts and enjoyment of our lives have come about as a direct result of science. Americans enjoy one of the highest living standards in the world due, to a large degree, to the accomplishments in agricultural science. Without a sound basis of agricultural knowledge, all humans would be struggling to find enough food, shelter, and clothing to survive. As a result of scientific research and the application of the findings of research, gigantic strides have been made in the efficient production of food from animals.

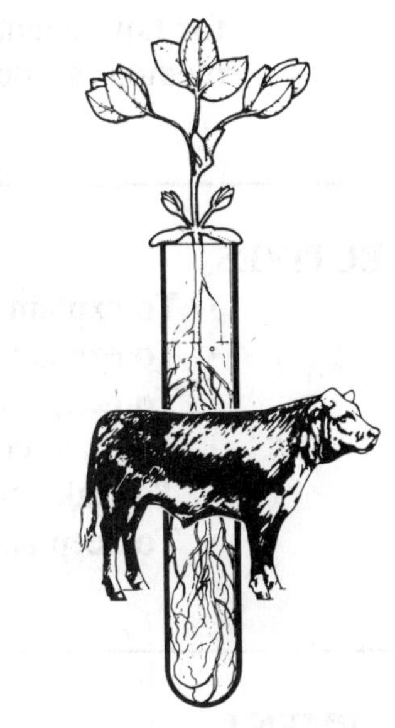

Figure 1.1.1

Researchers use a step-by-step procedure called the scientific method to solve problems and learn new information. Scientific study is a determined effort to look at a problem objectively and to learn the truth about a particular situation. Following a set procedure helps to ensure that the knowledge gained did not occur just by chance, or by the researcher's personal opinion, or by other factors. This process involves formulating a hypothesis, designing a study, collecting the data, and drawing conclusions based on an analysis of the data.

1. **Identify the problem:** A scientist begins by identifying the problem that needs to be solved. He or she may have an idea or suspicion of what causes the problem and what might solve the problem.

2. **Hypothesis:** The researcher uses his or her knowledge and observations about the problem to formulate a hypothesis. The hypothesis serves as the basis for investigating the problem.

3. **Design an experiment:** The hypothesis is then subjected to a test called an experiment that attempts to isolate the problem in question and to determine a solution.

4. **Collect and analyze data:** Careful measurements are taken and compiled for study.

5. **Draw conclusions:** The interpretation of the data is used to determine if the hypothesis is true. Statements that are supported by the data can then be made about the problem.

6. **Make recommendations:** Recommendations can then be made based on whether or not the hypothesis is true.

It is important for you to understand how the scientific method works — not because you are going to do research, but because you should have a feeling for how scientists carry out research and because you apply this same general method in your everyday life through problem solving (logical thinking).

OBJECTIVES

- To explain the importance of scientific research
- To explain the steps in the scientific method
- To design and conduct a simple scientific research study
- To draw conclusions from research data
- To make careful and accurate observations
- To keep accurate records of experimental results

EQUIPMENT

No special equipment is necessary for this exercise.

MATERIALS

No special materials are needed for this exercise.

SAFETY

There are no special safety precautions necessary for this exercise.

PROCEDURES

1. An experiment has been set up for you in Items 1 and 2 in the Results and Discussion section. This is a hypothetical situation to simulate the results you might get from a scientific study. In Item 1 you are to write the hypothesis and interpret the data given. In Item 2 you are to

make up and interpret data that will give a different result than that found in Item 1.

2. In Item 3 in the Results and Discussion section you are given a scenario (story outline) of a typical situation where research is needed. Follow through step-by-step to design a study to answer the question and draw conclusions.

RESULTS AND DISCUSSION

1. **SCENARIO:** A company is planning to market a new canned beef stew. The manager thinks that, in a stew, customers will not be able to tell the difference in USDA Choice, Select, or Standard grades of beef. The manager has to make a decision on which beef to use, but the financial risk is too great to make the decision on what she "thinks" to be true. The manager needs scientific data to back up her theory. The company can save money if the cheaper grades of beef (Select or Standard) can be used without any perceptible changes in quality and affecting sales.

   A scientific study has been designed to determine if customers like stews made from Select and Standard beef as well as they like stew made from the more expensive Choice beef. A researcher set up a station in a grocery store over several days and asked customers to taste the three stews. Customers were given small samples of the stews - labeled *A*, *B*, and *C* - and were asked which they liked best. All the stews were prepared in exactly the same manner with the same ingredients except for the grade of beef. Table 1.1.1 summarizes the data collected.

Statistical Significance

When you study Table 1.1.1 you will notice that the numbers 143 for Choice beef and 139 for Select beef are very close. You may wonder if there is enough difference to use one grade of beef over the other. In research this is called "statistical significance." In other words, is there really a difference or did these numbers happen just by chance? Some rather lengthy calculations can be made to answer this question. You will not be asked to make these calculations, but you will be asked to use your judgement in whether the numbers are significant. Example: 143 and 139 are essentially the same but 143 and 57 are certainly different enough that we can say they did not happen by chance.

What is the hypothesis of this study?

What conclusion should be drawn from the data in Table 1.1.1?

Table 1.1.1 Summary of Beef Grade Preference	
Beef stew made from:	Number who preferred A, B, or C
A. Choice beef	143
B. Select beef	139
C. Standard beef	57

2. If the data were different, different conclusions might be drawn. To illustrate this, fill in data in Table 1.1.2 that would give a different result from that found in Table 1.1.1. Use any numbers that will support the result you have chosen.

   What conclusions can be drawn from the data you inserted in Table 1.1.2?

Table 1.1.2 Summary of Beef Grade Preference	
Beef stew made from:	Number who preferred A, B, or C
A. Choice beef	
B. Select beef	
C. Standard beef	

3. Complete the details of the experiment outlined next to determine the best feed for finishing hogs.

SCENARIO: A researcher has heard from producers who say that fish meal is better than soybean meal as a protein supplement. Help design a study to determine if this is true.

RECOGNITION OF THE PROBLEM OR QUESTION: (Explain the question concerning fish meal and soybean meal.)

DEVELOP A HYPOTHESIS: (Based on the scenario, what do you think is true about the comparison of fish meal and soybean meal as a protein source for finishing hogs? Enter your hypothesis below.)

DESCRIBE THE EXPERIMENTAL DESIGN: (Help make sure the experiment tests what it is supposed to test by putting a check beside each item that should be done in this experiment.)

____ All the pigs should be about the same age.

____ All the pigs should be the same color.

____ There should be at least two groups.

____ There should be at least three pigs in each group to account for variability among individuals.

____ At the start of the test, the pigs should be about the same size.

____ The pigs should be assigned to groups at random.

____ The only difference in the treatment of the two groups should be the source of protein.

____ One group should be fed three times a day while the other group should be fed twice a day.

____ The pigs should all be weighed on the same day.

COLLECT DATA: Complete Table 1.1.3 by inserting data that will support your hypothesis (for illustration purposes you are being asked to "make up" data). Hint: Just so your data look reasonable, remember that finished market hogs weigh in the range of 200 - 250 pounds.

Table 1.1.3 Experimental Data Comparing Sources of Protein

Group 1 (soybean meal)				Group 2 (fish meal)			
	Starting Weight	Ending Weight	Weight Gained		Starting Weight	Ending Weight	Weight Gained
Pig 1	75			Pig 1	75		
Pig 2	76			Pig 2	76		
Pig 3	72			Pig 3	73		
Pig 4	71			Pig 4	69		
Total Weight Gained: ________				Total Weight Gained: ________			

ANALYZE THE DATA: The weight gained by the group fed fish meal will be compared to the group fed the soybean meal. How do they compare?

DRAW CONCLUSIONS: What are your conclusion(s) based on the data?

RECOMMENDATIONS: Assuming soybean meal and fish meal are the same price, what would you recommend to a producer as a source of protein for finishing hogs?

NAME:______________________ DATE:____________ CLASS:____________

Chapter 2: The Large Animal Industry

Laboratory Exercise 2.1 The Value of Large Animal Products

BACKGROUND

Americans are a nation of meat eaters. The United States ranks first in the world in the per capita consumption of poultry, third in beef and veal, and twenty-second in the consumption of pork.

The United States is well suited for the production of feed crops and the animals that supply meat. Throughout the midwestern portion of the United States millions of acres of corn are grown on some of the most productive farmland in the world. You might expect that this area also ranks high in the production of meat animals.

In this exercise you will use statistics from the USDA to determine the top producing states for beef, pork, and feed crops. You will also review your state's position in the production of agricultural animals and products relative to other states.

OBJECTIVES

- To compare the production of beef cattle and hogs among the states
- To compare the production of beef cattle and hogs in your state to other states
- To identify on a map the states and regions highest in the production of meat animals in the United States
- To determine relationships between feed crop production and the production of meat animals
- To determine the per capita consumption of pork, beef, and lamb and mutton

---

EQUIPMENT

No special equipment is needed for this exercise.

MATERIALS

References:

Ranking of States and Commodities by Cash Receipts (latest edition). Economics Research Service, United States Department of Agriculture, Herndon, VA 22070. Telephone 1-800-999-6779.

Statistical Abstract of the United States (latest edition). U.S. Department of Commerce, Bureau of the Census, Washington, DC 20402-9328.

SAFETY

No special safety precautions are necessary with this exercise.

PROCEDURES

1. Use the references to answer the items in the Results and Discussion section. Your teacher may direct you to use Tables 2.1.2 through 2.1.8 at the end of this exercise instead of the references listed.

RESULTS AND DISCUSSION

1. What is the per capita consumption of meat from cattle, hogs, and sheep in the United States? Complete Table 2.1.1 using the data found in the Statistical Abstract of the United States, Per Capita Consumption of Major Food Commodities. Draw arrows, up or down (↑ or ↓), to indicate if consumption has gone up or down from the first year reported to the last year reported.

2. How does the per capita consumption of meats compare with what you and your family eat?

   Do you and your family eat more or less

   Beef? ______________________________

   Veal? ______________________________

   Lamb and mutton? ___________________

   Pork? ______________________________

What are the possible reasons for the differences and/or similarities of the amount and kinds of meat your family eats as compared to the national average?

Table 2.1.1 Per Capita Consumption of Beef, Veal, and Lamb and Mutton

Commodity	Consumption Per Capita		
	Pounds 19___	Pounds 19___	Trend ↑ or ↓
Red meat, total			
Beef			
Veal			
Lamb and mutton			
Pork			

3. List the five leading states in the production of all meat animals.

   1.

   2.

   3.

   4.

   5.

4. The top five states produce ______ % of meat animals in the United States.

5. Compile the following statistics for your state:

   My state ranks __________ in the production of meat animals.

   The dollar value of meat animals in my state is ____________________ .

   My state produces ________ % of the total value of meat animals produced in the United States.

   The sale of meat animals represents ___________ % of the value of agricultural commodities in my state.

6. List the five leading states in the production of all livestock and livestock products.

   1.

   2.

   3.

   4.

   5.

7. The top five states produce ______ % of the value of all livestock and livestock products in the United States.

8. Compile the following statistics for your state.

My state ranks ________________ in the production of livestock and livestock products.

The dollar value of livestock and livestock products in my state is ________________ .

My state produces ________ % of the total value of livestock and livestock products produced in the United States.

The sale of livestock and livestock products represents ________ % of the value of agricultural commodities in my state.

9. List the five leading states in the production of cattle and calves.

1.

2.

3.

4.

5.

10. The top five states produce ______ % of the value of all cattle and calves produced in the United States.

11. Compile the following statistics for your state.

My state ranks ________ in the production of cattle and calves.

The dollar value of cattle and calves in my state is ________________ .

My state produces ________ % of the total value of cattle and calves produced in the United States.

The sale of cattle and calves represents ________ % of the value of agricultural commodities in my state.

12. List the five leading states in the production of hogs.

    1.

    2.

    3.

    4.

    5.

13. The top five states produce ______ % of the value of all hogs produced in the United States.

14. Compile the following statistics for your state:

    My state ranks ________________________ in the production of hogs.

    The dollar value of hogs in my state is ________________________ .

    My state produces ________ % of the total value of hogs produced in the United States.

    The sale of hogs represents _______ % of the value of agricultural commodities in my state.

15. Identify the top ten states in hogs and beef cattle by marking the map in Figure 2.1.1. Use the following code to identify the top ten states in the production of each animal: write the first letter of the animal and the rank of the state in the production of that animal. For example:

    B-6 means that the state ranks 6th in the production of beef.

    H-2 means that the state ranks 2d in hog production.

16. Identify the top ten states in feed crops by marking the map in Figure 2.1.1. Use the following code to identify the top ten states: Write *FC* to represent "feed crops" and a number to identify that state's rank in feed crop production. For example:

    FC-3 means that the state ranks 3d in feed crop production.

17. Is the production of feed crops related to the production of animals? Explain.

18. List and briefly discuss the factors that affect the numbers of large animals raised in your state.

19. If a state does not rank in the top ten in the production of meat animals, does that mean that meat animal production is relatively unimportant in that state? Explain your answer.

20. How do cattle and calves rank in value in the United States as compared to other commodities? ________________

21. How do hogs rank in value in the United States as compared to other commodities? ________________

22. How do sheep rank in value in the United States as compared to other commodities? ________________

23. Cattle, calves, and hogs together represent ____________% of the total United States receipts for farm commodities.

Table 2.1.2 Per Capita Consumption of Selected Food Commodities (pounds) 1970 - 1990

Commodity	1970	1975	1980	1985	1990
Red meat, total	132.0	125.3	126.4	124.9	112.3
Beef	79.6	83.0	72.1	74.6	64.0
Veal	2.0	2.8	1.3	1.5	.09
Lamb and Mutton	2.1	1.3	1.0	1.1	1.1
Pork	48.2	38.2	52.1	47.7	46.3
Fish and Shellfish	11.8	12.2	12.5	15.1	15.5
Poultry products, total	34.1	34.2	42.6	49.4	63.6
Chicken	27.7	27.5	34.3	39.9	49.3
Turkey	6.4	6.7	8.3	9.6	14.4
Eggs (number)	309	276	271	255	233
Milk, total (milk equivalent)	563.8	539.1	543.3	593.7	570.6
Fluid milk and cream	275.1	261.4	245.6	241.0	233.2
Milk, beverages	269.1	254.0	237.4	229.7	221.5
Cheese	11.4	14.3	17.5	22.5	24.7
Ice cream	17.8	18.6	17.5	18.1	15.7

Table 2.1.3 Meat Animals: States' Rankings for Cash Receipts, 1991

State and Rank		Value of commodity receipts	Percent of commodity total	Cumulative percent 1/	Percent of State's total for all commodities	State's total for all commodities
		1,000 dollars	..........................Percent..........................			1,000 dollars
1	Texas	6,316,421	12.3	12.3	52.0	12,126,182
2	Nebraska	5,669,847	11.1	23.4	64.2	8,821,328
3	Iowa	4,999,261	9.7	33.2	49.1	10,179,249
4	Kansas	4,606,470	9.0	42.2	66.4	6,934,986
5	Colorado	2,348,031	4.6	46.8	62.4	3,761,320
6	Oklahoma	2,278,294	4.4	51.3	59.8	3,807,582
7	Illinois	1,919,072	3.7	55.0	25.5	7,508,777
8	Minnesota	1,897,265	3.7	58.7	27.3	6,936,001
9	South Dakota	1,895,279	3.7	62.4	58.0	3,264,286
10	California	1,767,849	3.4	65.9	9.8	17,884,698
11	Missouri	1,412,249	2.7	68.7	36.5	3,861,179
12	Wisconsin	1,157,731	2.2	70.9	21.2	5,449,043
13	Indiana	1,131,551	2.2	73.2	25.2	4,474,513
14	Kentucky	915,181	1.7	74.9	28.7	3,178,704
15	North Carolina	835,331	1.6	76.6	16.9	4,924,071
16	Alabama	771,735	1.5	78.1	25.9	2,977,832
17	New Mexico	764,232	1.5	79.6	50.9	1,501,152
18	Ohio	750,217	1.4	81.1	19.2	3,893,074
19	Montana	721,578	1.4	82.5	47.1	1,531,169
20	Idaho	688,835	1.3	83.8	26.3	2,615,946
21	Arkansas	664,994	1.3	85.1	15.4	4,310,724
22	Wyoming	618,508	1.2	86.3	76.1	812,743
23	Pennsylvania	602,437	1.1	87.5	17.2	3,503,040
24	Tennessee	598,953	1.1	88.7	30.2	1,977,569
25	North Dakota	549,021	1.0	89.8	21.4	2,556,147
26	Arizona	545,652	1.0	90.8	28.8	1,889,907
27	Washington	543,432	1.0	91.9	13.7	3,946,524
28	Michigan	504,795	0.9	92.9	16.3	3,081,072
29	Virginia	502,708	0.9	93.9	23.9	2,095,371
30	Georgia	475,946	0.9	94.8	11.9	3,978,361
31	Oregon	472,875	0.9	95.7	19.2	2,454,389
32	Florida	380,832	0.7	96.5	6.2	6,140,999
33	Utah	301,682	0.5	97.0	41.2	730,882
34	Mississippi	276,949	0.5	97.6	11.4	2,422,070
35	Louisiana	227,577	0.4	98.0	12.6	1,792,907
36	New York	224,140	0.4	98.5	7.8	2,868,321
37	South Carolina	211,518	0.4	98.9	17.2	1,225,396
38	Nevada	145,487	0.2	99.2	52.7	275,836
39	West Virginia	113,616	0.2	99.4	34.4	330,237
40	Maryland	101,790	0.2	99.6	7.6	1,332,494
41	Vermont	46,955	0.0	99.7	10.8	433,140
42	Hawaii	39,494	0.0	99.8	6.6	596,925
43	Maine	27,810	0.0	99,8	6.2	444,601
44	New Jersey	17,445	0.0	99.9	2.6	660,160
45	Massachusetts	15,919	0.0	99.9	3.3	475,540
46	Connecticut	15,529	0.0	99.9	3.3	463,372
47	Delaware	10,076	0.0	99.9	61	9,536
48	New Hampshire	7,816	0.0	100.0	5.4	143,106
49	Rhode Island	1,730	0.0	100.0	2.4	70,917
50	Alaska	727	0.0	100.0	2.7	26,622
	United States	51,092,842			30.5	167,292,000

Numbers may not add due to rounding. 1/ The cumulative percentage is the sum of the percent of commodity total for each State and all preceding States.

Table 2.1.4 Livestock and Products: States' Rankings for Cash Receipts, 1991

State and Rank		Value of commodity group receipts	Percent of commodity group total	Cumulative percent 1/	Percent of State's total for all commodities	State's total for all commodities
		1,000 dollars	...Percent...			1,000 dollars
1	Texas	7,913,994	9.1	9.1	65.2	12,126,182
2	Nebraska	5,933,608	6.8	15.9	67.2	8,821,328
3	Iowa	5,720,916	6.6	22.5	56.2	10,179,249
4	California	5,272,141	6.0	28.6	29.4	17,886,698
5	Kansas	4,802,447	5.5	34.1	69.2	6,934,986
6	Wisconsin	4,215,045	4.8	39.0	77.3	5,449,043
7	Minnesota	3,576,920	4.1	43.1	51.5	6,936,001
8	Oklahoma	2,767,440	3.1	46.3	72.6	3,807,582
9	Arkansas	2,680,068	3.0	49.4	62.1	4,310,724
10	Colorado	2,663,835	3.0	52.5	70.8	3,761,320
11	North Carolina	2,608,480	3.0	55.5	52.9	4,924,071
12	Pennsylvania	2,469,760	2.8	58.3	70.5	3,503,040
13	Illinois	2,343,987	2.7	61.0	31.2	7,508,777
14	Alabama	2,219,173	2.5	63.6	74.5	2,977,832
15	Missouri	2,203,115	2.5	66.1	57.0	3,861,179
16	South Dakota	2,176,085	2.5	68.6	66.6	3,264,286
17	Georgia	2,153,489	2.4	71.1	54.1	3,978,361
18	Indiana	1,892,825	2.1	73.3	42.3	4,474,513
19	New York	1,781,592	2.0	75.3	62.1	2,868,321
20	Kentucky	1,703,977	1.9	77.3	53.6	3,178,704
21	Ohio	1,681,478	1.9	79.2	43.1	3,893,074
22	Virginia	1,362,945	1.5	80.8	65.0	2,095,371
23	Washington	1,289,910	1.4	82.3	32.6	3,946,524
24	Michigan	1,287,703	1.4	83.8	41.7	3,081,072
25	Mississippi	1,274,893	1.4	85.3	52.6	2,422,070
26	Florida	1,171,662	1.3	86.6	19.0	6,140,999
27	Idaho	1,072,779	1.2	87.8	41.0	2,615,946
28	Tennessee	1,045,055	1.2	89.0	52.8	1,977,569
29	New Mexico	1,019,388	1.1	90.2	67.9	1,501,152
30	Oregon	823,814	0.9	91.2	33.5	2,454,389
31	Montana	790,428	0.9	92.1	51.6	1,531,169
32	Arizona	785,637	0.9	93.0	41.5	1,889,907
33	Maryland	778,898	0.9	93.9	58.4	1,332,494
34	North Dakota	699,235	0.8	94.7	27.3	2,556,147
35	Wyoming	643,036	0.7	95.4	79.1	812,743
36	Louisiana	620,792	0.7	96.2	34.6	1,792,907
37	Utah	552,559	0.6	96.8	75.6	730,882
38	South Carolina	548,734	0.6	97.4	44.7	1,225,396
39	Delaware	438,065	0.5	97.9	70.7	619,536
40	Vermont	367,584	0.4	98.4	84.8	433,140
41	West Virginia	253,187	0.2	98.6	76.6	330,237
42	Maine	252,457	0.2	98.9	56.7	444,601
43	Connecticut	208,573	0.2	99.2	45.0	463,372
44	New Jersey	196,519	0.2	99.4	29.7	660,160
45	Nevada	187,090	0.2	99.6	67.8	275,836
46	Massachusetts	120,745	0.1	99.8	25.3	475,540
47	Hawaii	90,518	0.1	99.9	15.1	596,925
48	New Hampshire	62,944	0.0	99.9	43.9	143,106
49	Rhode Island	13,275	0.0	99.9	18.7	70,917
50	Alaska	6,478	0.0	100.0	24.3	26,622
	United States	86,745,278			51.8	167,292,000

Numbers may not add due to rounding. 1/ The cumulative percentage is the sum of the percent of commodity total for each State and all preceding States.

Table 2.1.5 Cattle and Calves: States' Rankings for Cash Receipts, 1991

State and Rank		Value of commodity receipts	Percent of commodity total	Cumulative percent 1/	Percent of State's total for all commodities	State's total for all commodities
		1,000 dollars	..........Percent..........			1,000 dollars
1	Texas	6,156,802	15.5	15.5	50.7	12,126,182
2	Nebraska	4,783,085	12.0	27.6	54.2	8,821,328
3	Kansas	4,271,740	10.7	38.3	61.6	6,934,986
4	Colorado	2,244,332	5.6	44.0	59.6	3,761,320
5	Oklahoma	2,225,970	5.6	49.6	58.4	3,807,582
6	Iowa	2,057,448	5.1	54.8	20.2	10,179,249
7	California	1,681,643	4.2	59.1	9.4	17,886,698
8	South Dakota	1,470,476	3.7	62.8	45.0	3,264,286
9	Minnesota	977,560	2.4	65.2	14.0	6,936,001
10	Wisconsin	923,995	2.3	67.6	16.9	5,449,043
11	Missouri	874,160	2.2	69.8	22.6	3,861,179
12	Kentucky	757,902	1.9	71.7	23.8	3,178,704
13	New Mexico	751,678	1.9	73.6	50.0	1,501,152
14	Illinois	739,857	1.8	75.4	9.8	7,508,777
15	Idaho	666,736	1.6	77.1	25.4	2,615,946
16	Montana	665,647	1.6	78.8	43.4	1,531,169
17	Alabama	645,505	1.6	80.4	21.6	2,977,832
18	Wyoming	590,249	1.4	81.9	72.6	812,743
19	Washington	533,669	1.3	83.3	13.5	3,946,524
20	Arkansas	519,525	1.3	84.6	12.0	4,310,724
21	Arizona	515,831	1.3	85.9	27.2	1,889,907
22	North Dakota	496,778	1.2	87.1	19.4	2,556,147
23	Tennessee	484,791	1.2	88.4	24.5	1,977,569
24	Pennsylvania	448,111	1.1	89.5	12.7	3,503,040
25	Oregon	443,769	1.1	90.6	18.0	2,454,389
26	Virginia	412,125	1.0	91.6	19.6	2,095,371
27	Florida	363,351	0.9	92.6	5.9	6,140,999
28	Ohio	325,205	0.8	93.4	8.3	3,893,074
29	Indiana	304,489	0.7	94.2	6.8	4.474.513
30	Michigan	284,605	0.7	94.9	9.2	3,081,072
31	Utah	283,178	0.7	95.6	38.7	730,882
32	Georgia	265,818	0.6	96.3	6.6	3,978,361
33	Mississippi	252,004	0.6	96.9	10.4	2,422,070
34	Louisiana	220,695	0.5	97.4	12.3	1,792,907
35	New York	207,574	0.5	98.0	7.2	2,868,321
36	North Carolina	176,396	0.4	98.4	3.5	4,924,071
37	Nevada	140,757	0.3	98.8	51.0	275,836
38	South Carolina	135,213	0.3	99.1	11.0	1,225,396
39	West Virginia	104,029	0.2	99.4	31.5	330,237
40	Maryland	73,736	0.1	99.6	5.5	1,332,494
41	Vermont	44,681	0.1	99.7	10.3	433,140
42	Hawaii	32,667	0.0	99.8	5.4	596,925
43	Maine	25,059	0.0	99.8	5.6	444,601
44	New Jersey	15,026	0.0	99.9	2.2	660,160
45	Connecticut	14,116	0.0	99.9	3.0	463,372
46	Massachusetts	11,341	0.0	99.9	2.3	475,540
47	New Hampshire	6,503	0.0	99.9	4.5	143,106
48	Delaware	4,596	0.0	100.0	0.7	619,536
49	Rhode Island	1,087	0.0	100.0	1.5	70,917
50	Alaska	578	0.0	100.0	2.1	26,622
	United States	39,632,088			23.6	167,292,000

Numbers may not add due to rounding. 1/ The cumulative percentage is the sum of the percent of commodity total for each State and all preceding States.

Table 2.1.6 Hogs: States' Rankings for Cash Receipts, 1991

State and Rank		Value of commodity receipts	Percent of commodity total	Cumulative percent 1/	Percent of State's total for all commodities	State's total for all commodities
		1,000 dollars	............................Percent	....		1,000 dollars
1	Iowa	2,916,499	26.3	26.3	28.6	10,179,249
2	Illinois	1,172,411	10.6	36.9	15.6	7,508,777
3	Minnesota	906,539	8.2	45.1	13.0	6,936,001
4	Nebraska	878,134	7.9	53.1	9.9	8,821,328
5	Indiana	823,759	7.4	60.5	18.4	4,474,513
6	North Carolina	658,649	5.9	66.5	13.3	4,924,071
7	Missouri	534,253	4.8	71.3	13.8	3,861,179
8	Ohio	415,360	3.7	75.0	10.6	3,893,074
9	South Dakota	396,845	3.5	78.6	12.1	3,264,286
10	Kansas	322.798	2.9	81.5	4.6	6,934,986
11	Wisconsin	229,558	2.0	83.6	4.2	5,449,043
12	Michigan	216,234	1.9	85.6	7.0	3,081,072
13	Georgia	209,496	1.8	87.5	5.2	3,978,361
14	Kentucky	156,200	1.4	88.9	4.9	3,178,704
15	Pennsylvania	150,335	1.3	90.2	4.2	3,503,040
16	Arkansas	145,049	1.3	91.6	3.3	4,310,724
17	Alabama	126,032	1.1	92.7	4.2	2,977,832
18	Tennessee	113,768	1.0	93.7	5.7	1,977,569
19	Texas	95,155	0.8	94.6	0.7	12,126,182
20	Virginia	85,204	0.7	95.4	4.0	2,095,371
21	South Carolina	76,268	0.6	96.0	6.2	1,225,396
22	Colorado	68,241	0.6	96.7	1.8	3,761,320
23	Oklahoma	49,039	0.4	97.1	1.2	3,807,582
24	California	45,224	0.4	97.5	0.2	17,886,698
25	North Dakota	44,952	0.4	97.9	1.7	2,556,147
26	Montana	36,845	0.3	98.3	2.4	1,531,169
27	Maryland	27,227	0.2	98.5	2.0	1,332,494
28	Mississippi	24,738	0.2	98.7	1.0	2,422,070
29	Arizona	20,939	0.1	98.9	1.1	1,889,907
30	Florida	17,304	0.1	99.1	0.2	6,140,999
31	New York	14,529	0.1	99.2	0.5	2,868,321
32	Oregon	14,135	0.1	99.3	0.5	2,454,389
33	Idaho	10,558	0.1	99.4	0.4	2,615,946
34	West Virginia	7,234	0.0	99.5	2.1	330,237
35	Washington	6,802	0.0	99.5	0.1	3,946,524
36	Louisiana	6,650	0.0	99.6	0.3	1,792,907
37	Hawaii	6,463	0.0	99.7	1.0	596,925
38	Delaware	5,440	0.0	99.7	0.8	619,536
39	Utah	4,931	0.0	99.8	0.6	730,882
40	New Mexico	4,419	0.0	99.8	0.2	1,501,152
41	Wyoming	4,379	0.0	99.8	0.5	812,743
42	Massachusetts	3,653	0.0	99.9	0.7	475,540
43	Nevada	2,493	0.0	99.9	0.9	275,836
44	Maine	1,952	0.0	99.9	0.4	444,601
45	New Jersey	1,288	0.0	99.9	0.2	660,160
46	Vermont	959	0.0	99.9	0.2	433,140
47	Connecticut	957	0.0	99.9	0.2	463,372
48	New Hampshire	844	0.0	99.9	0.5	143,106
49	Rhode Island	576	0.0	100.0	0.8	70,917
50	Alaska	124	0.0	100.0	0.4	26,622
	United States	11,061,441			6.6	167,292,000

Numbers may not add due to rounding. 1/ The cumulative percentage is the sum of the percent of commodity total for each state and all preceding states.

Table 2.1.7 Feed Crops: States' Rankings for Cash Receipts, 1991

State and Rank		Value of commodity receipts	Percent of commodity total	Cumulative percent 1/	Percent of State's total for all commodities	State's total for all commodities
		1,000 dollars	....................Percent	....		1,000 dollars
1	Illinois	2,755,737	14.4	14.4	36.7	7,508,777
2	Iowa	2,563,247	13.4	27.9	25.1	10,179,249
3	Nebraska	2,051,387	10.7	38.7	23.2	8,821,328
4	Minnesota	1,429,772	7.5	46.2	20.6	6,936,001
5	Indiana	1,269,783	6.6	52.9	28.3	4,474,513
6	Texas	897,419	4.7	57.6	7.4	12,126,182
7	Ohio	841,835	4.4	62.1	21.6	3,893,074
8	Kansas	822,350	4.3	66.4	11.8	6,934,986
9	California	616,654	3.2	69.6	3.4	17,886,698
10	Missouri	512,272	2.6	72.3	13.2	3,861,179
11	Colorado	485,747	2.5	74.9	12.9	3,761,320
12	Michigan	485,206	2.5	77.4	15.7	3,081,072
13	Wisconsin	462,322	2.4	79.9	8.4	5,449,043
14	South Dakota	401.074	2.1	82.0	12.2	3,264,286
15	North Dakota	344,965	1.8	83.8	13.5	2,556,147
16	Idaho	278,956	1.4	85.3	10.6	2,615,946
17	Kentucky	274,135	1,4	86,7	8,6	3,178,704
18	Washington	251,921	1.3	88.0	6.3	3,946,524
19	Montana	233,506	1.2	89.3	15.2	1,531,169
20	North Carolina	188,556	0.9	90.2	3.8	4,924,071
21	Pennsylvania	166,463	0.8	91.1	4.7	3,503,040
22	New Mexico	153.236	0.8	91.9	10.2	1,501,152
23	New York	149,436	0.7	92.7	5.2	2,868,321
24	Oregon	136,962	0.7	93.4	5.5	2,454,389
25	Georgia	134,805	0.7	94.1	3.3	3,978,361
26	Arizona	113,931	0.6	94.7	6.0	1,889,907
27	Oklahoma	104,704	0.5	95.3	2.7	3,807,582
28	Maryland	103,286	0.5	95.8	7.7	1,332,494
29	Tennessee	97,006	0.5	96.3	4.9	1,977,569
30	Arkansas	81,802	0.4	96.8	1.9	4,310,724
31	Wyoming	77,367	0.4	97.2	9.5	812,743
32	Virginia	77,143	0.4	97.6	3.6	2,095,371
33	Utah	69,995	0.3	98.0	9.5	730,882
34	Louisiana	67,076	0.3	98.3	3.7	1,792,907
35	Nevada	49,089	0.2	98.6	17.8	275,836
36	South Carolina	48,146	0.2	98.8	3.9	1,225,396
37	Delaware	41,396	0.2	99.0	6.6	619,536
38	Alabama	39,019	0.2	99.2	1.3	2,977,832
39	Mississippi	35,959	0.1	99.4	1.4	2,422,070
40	New Jersey	24,542	0.1	99.6	3.7	660,160
41	Florida	17,326	0.0	99.7	0.2	6,140,999
42	West Virginia	15,618	0.0	99.7	4.7	330,237
43	Vermont	13,149	0.0	99.8	3.0	433,140
44	Connecticut	7,912	0.0	99.8	1.7	463,372
45	New Hampshire	6,368	0.0	99.9	4.4	143,106
46	Maine	6,047	0.0	99.9	1.3	444,601
47	Massachusetts	4,689	0.0	99.9	0.9	475,540
48	Alaska	2,472	0.0	100.0	9.2	26,622
49	Rhode Island	588	0.0	100.0	0.8	70,917
	United States	19,012,376			11.3	167,292,000

Numbers may not add due to rounding. 1/ The cumulative percentage is the sum of the percent of commodity total for each State and all preceding States.

Table 2.1.8 United States: Leading Commodities for Cash Receipts, 1991

Item and Rank		Value of U.S. receipts	Percent of U.S. total	Cumulative percent 1/	Rank in prior year
		1,000 dollars	...............Percent...........		
	All commodities	167,292,000	100.0		
	Livestock and Products	86,745,278	51.8		
	Crops	80,546,722	48.1		
1	Cattle and calves	39,632,088	23.6	23.6	1
2	Dairy products	18,113,714	10.8	34.5	2
3	Corn	13,853,798	8.2	42.8	3
4	Hogs	11,061,441	6.6	49.4	4
5	Soybeans	10,778,421	6.4	55.8	5
6	Greenhouse and nursery	8,404,722	5.0	60.8	7
7	Broilers	8,385,284	5.0	65.8	6
8	Wheat	5,715,687	3.4	69.3	8
9	Cotton	5,588,934	3.3	72.6	9
10	Chicken eggs	3,861,358	2.3	74.9	10
11	Hay	3,044,049	1,8	76.7	11
12	Tobacco	2,886,039	1.7	78.5	12
13	Turkeys	2,344,016	1.4	79.9	14
14	Potatoes	2,047,785	1.2	81.1	13
15	Tomatoes	1,798.806	1.0	82.2	17
16	Apples	1,659,334	0.9	83.1	20
17	Grapes	1,618,558	0.9	84.1	16
18	Oranges	1,564,762	0.9	85.1	15
19	Peanuts	1,392,537	0.8	85.9	18
20	Sugar beets	1,216,758	0.7	86.6	19
21	Sorghum grain	1,131,780	0.6	87.3	22
22	Rice	1,092,385	0.6	87.9	21
23	Cane for sugar	902,031	0.5	88.5	24
24	Barley	821,519	0.4	89.0	25
25	Lettuce	817,667	0.4	89.5	23
Government payments 2/		8,214,399			

Numbers may not add due to rounding. 1/ The cumulative percentage is the sum of the percent of U.S. total for each commodity and all preceding commodities.
2/ Government payment made directly to farmers in cash or Payment-in-Kind.

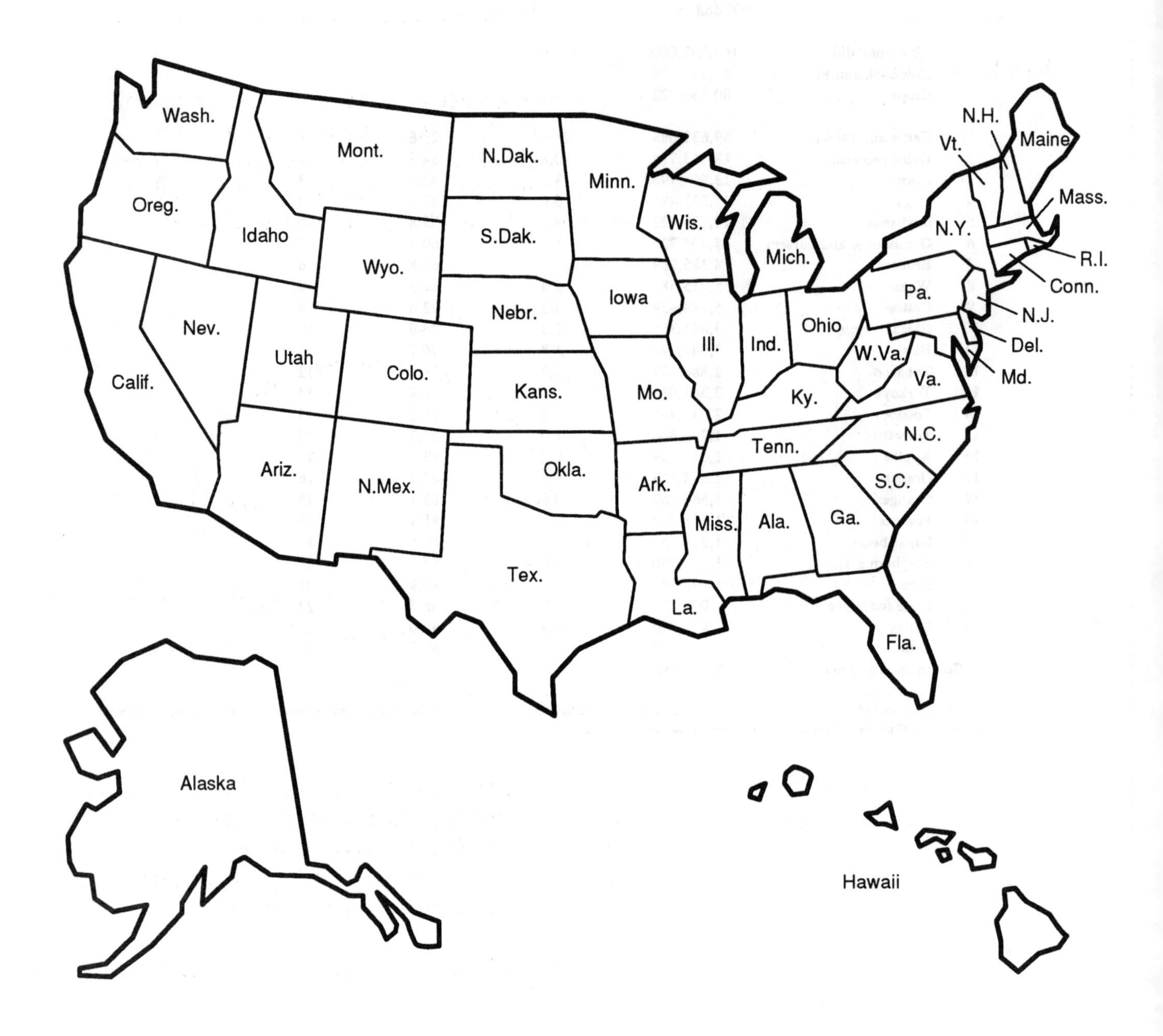

Wash.
Mont.
N.Dak.
Minn.
Oreg.
Idaho
Wyo.
S.Dak.
Wis.
Mich.
N.Y.
Vt.
N.H.
Maine
Mass.
R.I.
Conn.
N.J.
Del.
Md.
Calif.
Nev.
Utah
Colo.
Nebr.
Iowa
Ill.
Ind.
Ohio
Pa.
W.Va.
Va.
Kans.
Mo.
Ky.
N.C.
Tenn.
Ariz.
N.Mex.
Okla.
Ark.
S.C.
Miss.
Ala.
Ga.
Tex.
La.
Fla.
Alaska
Hawaii

Figure 2.1.1

NAME:____________ DATE:__________ CLASS:__________

Chapter 3: The Poultry Industry

Laboratory Exercise 3.1 The Parts of an Egg

BACKGROUND

The pores of an egg shell allow air exchange between the egg or developing embryo and the outside environment. Anything that clogs the pores or otherwise prevents air exchange could be detrimental to the developing chick. Just like other animals, the chick embryo uses oxygen and gives off carbon dioxide. In Part I of this exercise you will observe the release of air through the pores of an egg.

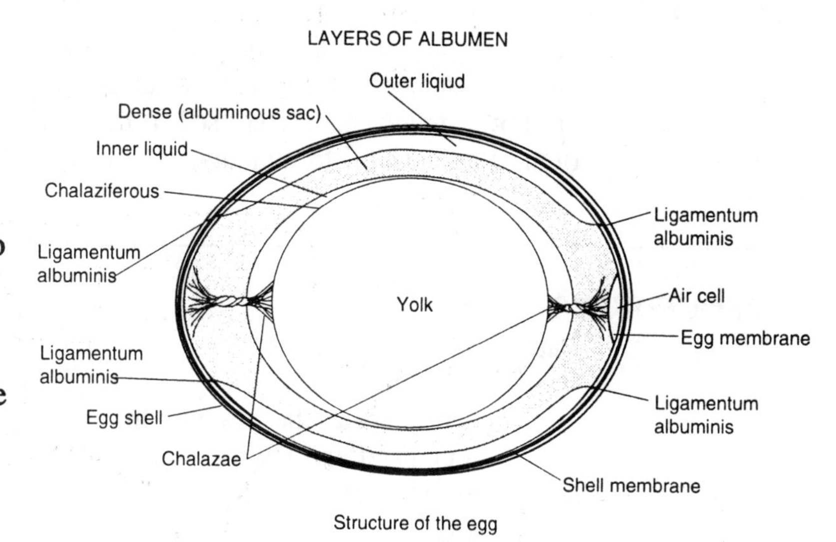

Figure 3.1.1 A cross section of an egg

Most people recognize only the yoke and white of the egg. Sometimes, when a person takes time to closely examine a raw egg, they see various parts that they think should be removed before cooking. For example, some people try to remove the chalazae. The chalazae is a stringy-like white substance that helps hold the yoke in the center of the egg. There is no reason to remove the chalazae or other parts of healthy eggs. Sometimes blood spots are found. Blood spots are certainly undesirable, but compared to the amount of blood we eat in pork, chicken, and beef, the tiny speck sometimes found in an egg is inconsequential.

In this exercise you will also learn to identify the parts of the egg and the role they play in the development of a chick.

OBJECTIVES

- To determine how an egg or developing chick embryo breathes
- To identify the parts of the egg

EQUIPMENT

hot plate or hot water source
400-ml beaker
saucer

MATERIALS

one raw egg
plastic spoon or fork

SAFETY

Avoid contact with the hot plate.
Do not heat the water beyond a temperature comfortable to the touch.
Do not taste any materials used in this lab. Raw eggs can carry disease organisms harmful to humans.

PROCEDURES

Part I

1. Gently shake the raw egg in your hand and try to detect any movement within the egg.

2. Place the egg in the beaker or pan with enough warm water to cover the egg. Place the container on a hot plate and heat the water slowly until it reaches a temperature still bearable to the touch.

Figure 3.1.2

3. Watch the egg for bubbles rising from it. (See Figure 3.1.2.) Sketch the egg in the space provided in the Results and Discussion section. Mark the pointed end and blunt ends of the egg. Mark on your sketch the spots where bubbles are escaping.

Part II

4. Remove the egg from the water and gently crack it over a saucer. Carefully pour the contents into the saucer.

5. Examine the insides of the shell. Locate the blunt end of the egg and examine the airspace located on the inside of the shell.

6. Sketch the broken egg and its parts in the space provided in #3 of the Results and Discussion section. Label the parts by referring to the diagram on page 45 of the text.

7. Find and label each of the following parts: germinal disk, chalazae, albumen, and yoke. You may have to gently turn the yolk with the spoon to locate the germinal disk.

RESULTS AND DISCUSSION

1. Why could you not feel the contents of the egg moving as you shook it back and forth?

2. Sketch the egg and the bubbles rising from it as indicated in step 3 in the Procedures section.

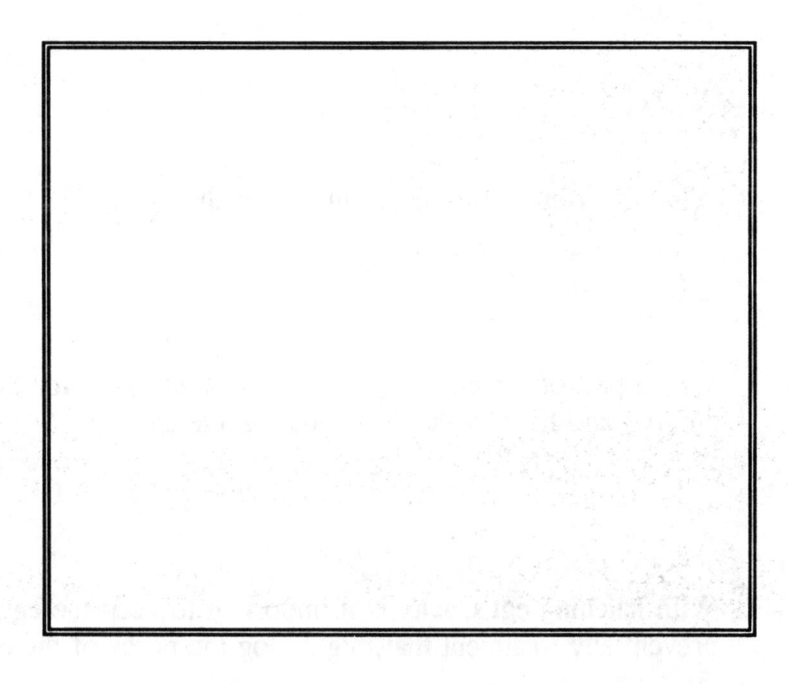

3. Sketch the broken egg below and label the parts.

4. Where were most of the bubbles coming from when you placed the egg in warm water? Why?

5. Based on what you observed in Part I, how does a developing embryo get air?

6. What did you find in the blunt end of the egg shell?

7. Which part of the egg serves as a shock absorber for the developing embryo and helps to hold the yoke in the center?

8. With hatching eggs, why is it important to keep the eggs clean and to prevent any treatment that might clog the pores of the egg shell?

NAME:____________________ DATE:____________ CLASS:____________

Chapter 3 : The Poultry Industry

Laboratory Exercise 3.2 Determining the Freshness of Eggs

BACKGROUND

Like most agricultural products, eggs are perishable. The major factor in the egg freshness is the age of the egg, but the storage conditions they receive between the farm and consumption are also very important factors. The quality of eggs cannot be improved upon - only preserved.

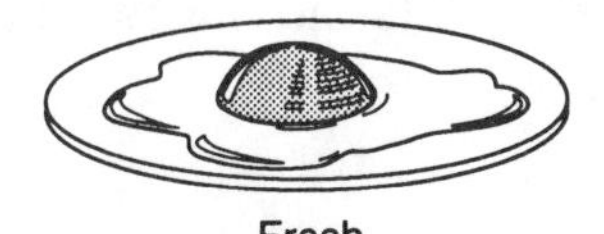

Fresh Not fresh

To determine the freshness of an egg, break it onto a plate. If the yolk stands high, the egg is fresh, If it breaks and runs into the white, the egg is safe to eat but past its prime.

Figure 3.2.1 The freshness of an egg can be estimated by the height of the yoke

How long will an egg keep? That is a difficult question to answer because it depends on many conditions. Producers gather eggs on a daily basis. They wash, cool, and sometimes oil the eggs to keep them fresher. The processor takes great pains to see that the humidity and temperature are correct. They also candle the eggs to detect blood and meat spots or other problems. Here is where things can go wrong. If eggs are hauled long distances without refrigeration, if the fast food outlet personnel sits them out in a hot kitchen, or the grocer fails to maintain proper temperature, the quality can deteriorate rapidly.

Eggs should be stored at 40^0 - 55^0 F for best keeping quality. Fluctuations in temperature or long periods out of refrigeration will decrease the quality. If eggs are removed from refrigeration, they may begin to "sweat." When they are again refrigerated, some of the water that has accumulated on the shell may be pulled back into the egg through the pores as the center of the egg begins to shrink or contract as it cools. Bacteria and other disease-causing organisms on the surface of the egg may also be pulled inside the shell.

OBJECTIVES

- To identify factors contributing to spoilage of eggs
- To determine proper storage conditions for eggs

EQUIPMENT

one glass bowl or pan at least six inches tall
two saucers or petri dishes
ruler

MATERIALS

Eggs of varying degrees of freshness

SAFETY

Do not taste any materials used in this lab. Raw eggs can carry disease organisms harmful to humans.

PROCEDURES

1. Obtain a sample of eggs of varying freshness from the teacher. The teacher should have at least two samples of eggs of various degrees of freshness. The teacher may have numbered the eggs or may direct you to number the eggs so that after the exercise it can be determined from which batch they came.

2. It will be your objective to determine the freshness of the eggs assigned to you and to determine the freshness of the batches from which they were taken.

3. Fill a large glass or pan with a column of water at least three times the height of an egg (about 6").

4. Number the eggs sequentially with a pencil and gently place them in the water. Use only one egg at a time unless your container is large enough so that more than one at a time can move freely.

5. Determine if the eggs sink or float and the degree to which they float. Record your findings in the Results and Discussion section. Use a ruler to measure the depth to which the egg sinks.

6. Select the two eggs that sink the deepest and float the highest. Crack them into separate saucers and visually inspect them for signs of freshness.

7. View the two eggs from the side, as shown in Figure 3.2.1.

8. Look for differences in the shapes of the two yolks.

RESULTS AND DISCUSSION

1. Record your findings on the freshness of the eggs below. Measure how far each egg floats from the bottom of the container.

Table 3.2.1 Determining the Freshness of Eggs

Egg Sample Number	To what degree did the egg float? (Measure in inches from the bottom).	Is this egg fresh or spoiled?
1		
2		
3		
4		

2. Fresh eggs __________ (sink or float) and spoiled eggs ________ (sink or float). How do you account for the change in buoyancy?

3. Viewing the broken eggs from the sides, is one yolk more rounded and taller than the other? Explain. (Compare the shape of the two yolks with the drawings in Figure 3.2.1)

NAME:______________________ DATE:__________ CLASS:__________

Chapter 3: The Poultry Industry

Laboratory Exercise 3.3 Egg Shell Strength

BACKGROUND

Egg grades are based on several factors, one of which is cracks. Cracked eggs pose a threat to the safety of the consumer and cause financial loss to the producers and processors. Disease-causing organisms may enter through cracks in the shell. Under no circumstances should a cracked egg be used in a recipe that calls for eggs to be consumed raw such as in egg nog. Cracked eggs may be used in cooked recipes under proper conditions.

Mother nature has designed the egg well for survival. The egg shell is very strong, especially on the ends. The dome shape is one of the strongest designs, as you will see in the following exercise. Dome structures are used extensively in architecture for strength as well as beauty.

The egg is shaped so that it rolls in a circle - perhaps so that it will not roll far from the nest and become damaged or lost. Eggs have a small end and a blunt end, which make them fit well into a nest.

OBJECTIVES

- To determine the strength of egg shells
- To recognize the health hazards and economic loss of cracked eggs
- To design a method of protecting eggs during shipment
- To determine the best position of eggs during shipment

EQUIPMENT

laboratory scale
books of various thicknesses

MATERIALS

one egg
two Ziploc® type sandwich bags
paper towel

SAFETY

Do not taste any materials used in this exercise. Raw eggs can carry disease organisms that are harmful to humans.

PROCEDURES

Part I

1. Select an egg and inspect it for cracks. Only unblemished eggs should be used in this exercise.

2. Place the egg inside a sandwich bag, remove all the air from the bag, and zip it closed.

3. Place the egg between your hands with one end in each palm, as shown in Figure 3.3.1. With your fingers laced together, squeeze the egg, being careful to put force only on the ends of the egg. If force is concentrated properly on the ends, you will not be able to break the egg.

Figure 3.3.1

Part II

4. Place the egg in the sandwich bag on the floor and have one student gently balance it on end. A paper towel placed inside the bag may help balance the egg. The person holding the egg must be very careful not to exert any unnecessary pressure on the side of the egg.

5. Have a student begin to place books on the end of the egg while another student balances the books without exerting any force other than that necessary to keep the books level. (See Figure 3.3.2.)

6. Continue to place books on the stack until the egg is crushed. Weigh the books on the scale to determine the amount of force necessary to break the egg.

Part III

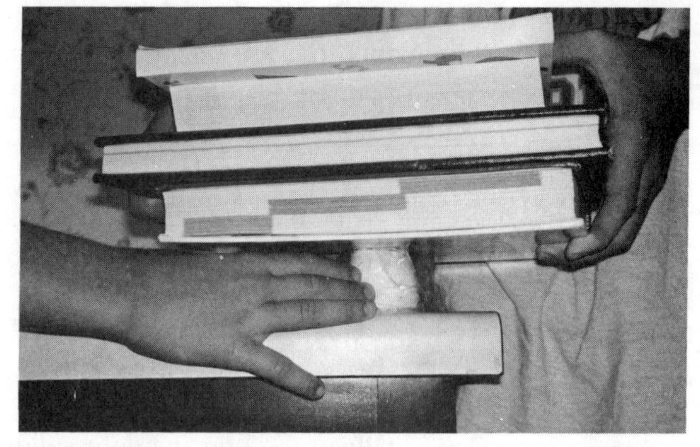
Figure 3.3.2

7. Design a package that will protect an egg from breaking when dropped from a height specified by your teacher. The package may be made from a variety of materials. The egg may suspended or held inside the container in any manner as long as it is secure and remains (broken or unbroken) inside the crash container after dropping.

8. Prior to securing the egg in your crash container, be sure to enclose the egg in a Ziploc® bag with the air removed.

9. Drop the containers from increasing heights to determine a winning design. Compare the successful designs in the class.

RESULTS AND DISCUSSION

1. Why didn't the egg crack when you pressed the ends between your hands?

2. How is the principle illustrated in item 1 above used in architecture?

3. Explain the design of an egg carton based on the results of this experiment.

4. How many pounds of books did it take to crack the egg?

5. Based on your findings, what do you think might be the maximum weight of a hen sitting on eight eggs before she crushed them?

6. Why is proper feeding of hens to ensure a strong egg shell important to layer operations and people who market eggs?

7. What type of container protects eggs best from dropping? Why was the design successful?

NAME:______________________________ DATE:______________ CLASS:______________

Chapter 3: The Poultry Industry

Laboratory Exercise 3.4 Measuring Relative Humidity for Egg Storage

---

BACKGROUND

The care and handling of hatching eggs is of upmost importance in the hatching rate. The two most important factors are temperature and humidity. Research has shown that hatchability is best preserved at temperatures between 55° and 65°F and a relative humidity of about 70%.

This exercise will deal with measuring and controlling relative humidity. Relative humidity is the percent of moisture in the air compared to the amount it can hold at that temperature. If the relative humidity is 85%, that means that the air contains 85% of the moisture that it possibly can hold at that temperature. The temperature is important because warm air holds more moisture than cool air. For example, if classroom air at 70°F with a 50% humidity level is brought into the incubator and raised to the hatching temperature of 100°F, the percent humidity will drop. The resulting relative humidity will be too dry for egg storage or hatching.

Eggs are about 60% water. Water is essential in the development of the embryo. Under dry air conditions (low humidity), water will gradually evaporate through pores in the shell. Proper hatching conditions require that the eggs be stored and hatched under high-humidity conditions.

In this exercise you will learn to calculate relative humidity. Your school may have a wet-bulb/dry-bulb hygrometer, which is a scientifically prepared device for measuring humidity. You may want to use the hygrometer to test your homemade device. Other hygrometers are made of human hair - an especially sensitive material to changes in humidity as any curly haired person can tell you on a humid day. Electrical hygrometers measure changes in electrical resistance of a water absorbing strip.

OBJECTIVES

- To determine the proper storage conditions of hatching eggs
- To measure and or calculate relative humidity

EQUIPMENT

two thermometers

electric fan (optional)
commercially made hygrometer (optional)

MATERIALS

one two-inch-square piece of absorbent cloth
psychometric chart
rubber band
a cup of room temperature water

SAFETY Avoid breakage of the thermometers.
Keep the wet-bulb thermometer at least six inches from the electric fan.

PROCEDURES

1. Place a small piece (1" - 2" square) of absorbent cloth around the tip of one thermometer and secure it with a string or rubber band. This thermometer will be used to get the wet bulb reading.

2. Place the other thermometer (the dry bulb thermometer) in a position so that it will accurately measure the air temperature in the room. Avoid drafts, sunny windows, heaters, and air conditioners.

3. Dip the wet bulb thermometer in room temperature water and hold it in front of a fan. If a fan is not available, carefully swing the thermometer in the air to aid air circulation and evaporation.

4. After five minutes take readings from both the wet bulb and dry bulb thermometer. Record the measurements in the Results and Discussion section.

5. Use the two thermometer readings to calculate the relative humidity from the psychrometric chart in Figure 3.4.1. Record your measurements and calculations in Table 3.4.2 and complete the Results and Discussion section. Complete this procedure twice to confirm your results.

6. Directions for reading a psychrometric chart.

   a. Find the dry bulb temperature on the vertical lines at the bottom of the chart.

b. Find the wet bulb temperature on the diagonal lines on the left side of the chart.

c. Follow the diagonal line from the wet bulb reading to where it crosses the vertical line of the dry bulb reading.

d. Read the relative humidity from the curved lines marked with percent at the point where the dry bulb and wet bulb lines cross.

RESULTS AND DISCUSSION

1. Why is humidity so important to the development of a chick embryo?

2. What is the definition of *relative humidity*?

3. Why is the wet bulb reading always cooler than the dry bulb reading?

4. Complete Table 3.4.1 by calculating the relative humidity of the four examples.

Table 3.4.1 Examples of Relative Humidity Calculations

Example #	Dry Bulb Reading	Wet Bulb Reading	Relative Humidity
1	82	78	
2	90	80	
3	88	81	
4	94	87	

5. The wet bulb reading is the temperature the air in the room would be if the humidity was ______%.

6. When there is little difference in the dry bulb and wet bulb readings, would that indicate high or low humidity?

7. Record your readings and calculations in Table 3.4.2.

Table 3.4.2 Calculating Relative Humidity

Test Number	Dry Bulb Reading	Wet Bulb Reading	Degree Difference	Relative Humidity
1				
2				

8. If it was a rainy, overcast day and the room felt sticky and damp, what results would you expect? Why?

9. How often do you think humidity should be measured around hatching eggs?

10. Do you think humidity is important to keeping eggs for eating purposes? Why?

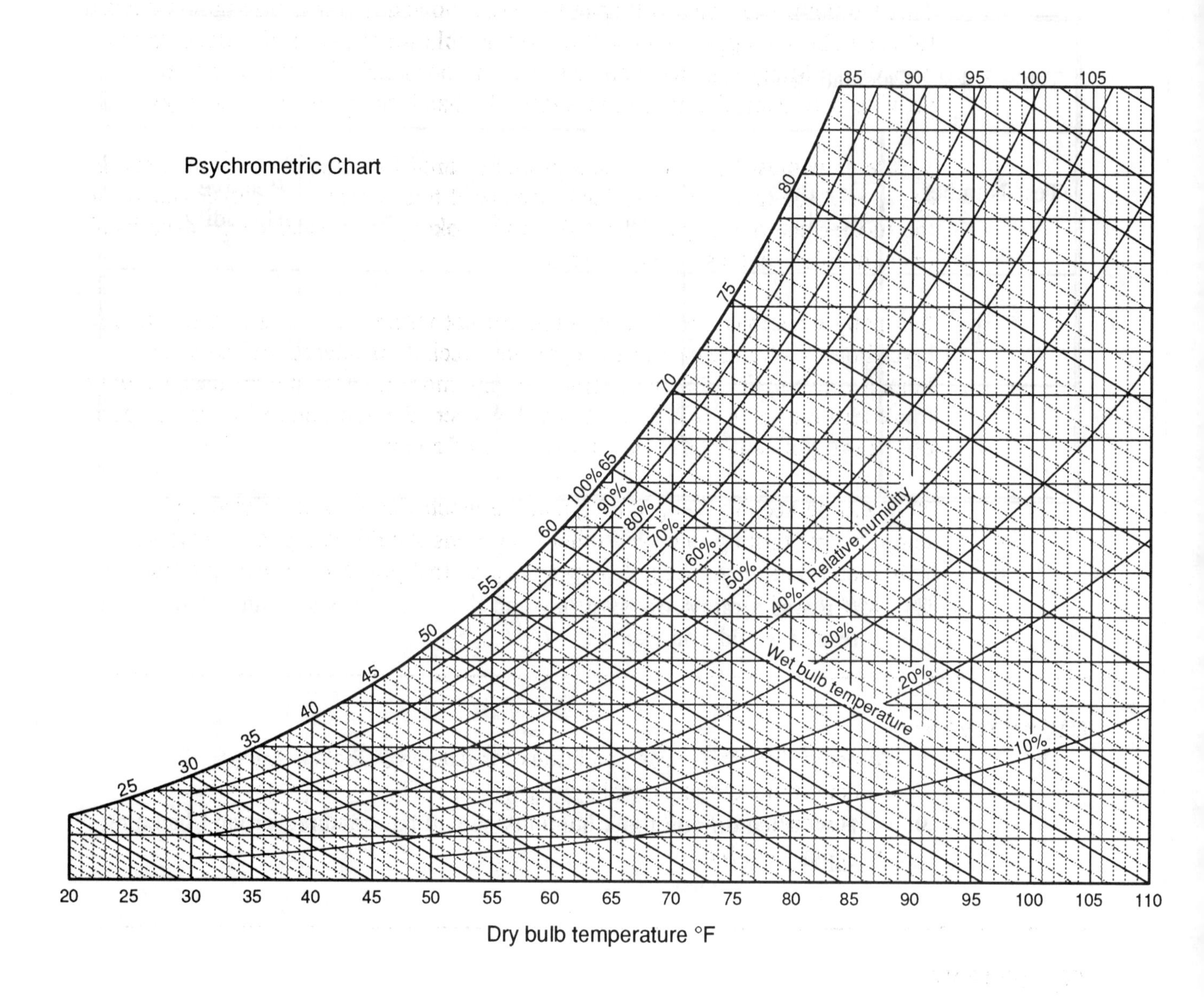

Figure 3.4.1 Psychrometric chart

NAME:______________________ DATE:____________ CLASS:____________

Chapter 3: The Poultry Industry

Laboratory Exercise 3.5 Hatching Chicks

BACKGROUND

This laboratory exercise will span the twenty-one day incubation period needed to hatch chicken eggs. You will be responsible for the care of a delicate living organism. Many eggs have defects and will not hatch, but the care you provide will determine to a large extent the hatchability of the good eggs.

Fertile eggs must be obtained and properly cared for until incubation. Fertile eggs are those from hens that have been bred to a rooster. A sperm cell from the rooster enters the germinal disk on the yoke. Your teacher will have local or mail-order sources of fertile eggs.

Normally eggs from the grocery store are not fertile. There are no roosters in layer houses. Sometimes fertile eggs are specially produced and sold for human consumption. Some people will pay more for fertile eggs because they think fertile eggs are more nutritious; however, the addition of one male sperm cell is insignificant to the nutritional value of an egg.

You will be keeping a log throughout the incubation period. Humidity, temperature, turning rate, and other conditions should be logged. Also you will keep a log of the weight of your egg so that you can answer questions in the Results and Discussion section.

OBJECTIVES

- To determine the proper conditions for hatching chicken eggs
- To determine and explain the weight gain or loss during incubation
- To demonstrate the use of a thermometer, thermostat, humidistat, incubator, and scale

---

EQUIPMENT

laboratory scale
egg incubator
shallow pan for water
thermometer
mechanical egg turner (optional)
egg candler

MATERIALS

one or more fertile chicken eggs

SAFETY

There are no special safety concerns in this exercise.

PROCEDURES

1. Obtain a fertile chicken egg and mark it for easy identification. Use a pencil or water-soluble pen so as not to block the pores in the shell. Include the egg number (if assigned by the teacher), the date, your name, and you may want to give the unborn chick a name. Mark an *X* on the egg to use as a reference point in turning the eggs. This will assist in making sure all the eggs are turned.

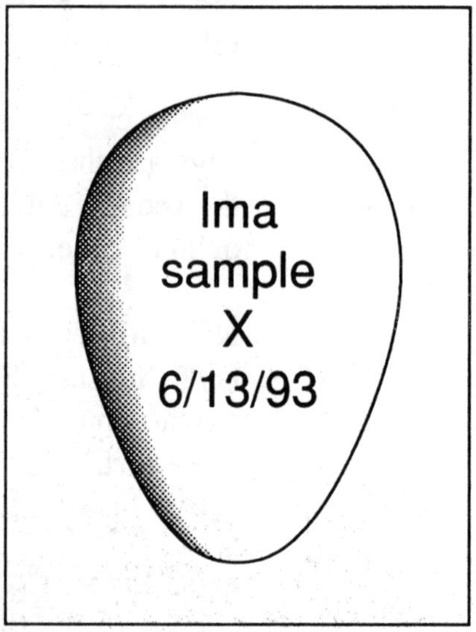

Ima
sample
X
6/13/93

Figure 3.5.1 Mark your egg with your name, date, and an "X"

2. Weigh your egg and record the weight in the chart in Table 3.5.1 in the Results and Discussion section. Handle the egg with great care! Weigh the egg each day through day seventeen and record the weight in the appropriate chart. This information will help you check your answers to some of the Results and Discussion questions that ask if you think the weight of the egg will change, and if so, will it lose weight or gain weight.

3. It is a good idea to calibrate the incubator before placing eggs in it to be hatched. Set the thermostat at 100°F. Leave the thermometer inside the incubator until you are sure the thermostat is accurate. You may also need to check your thermometer for accuracy. When the temperature has stabilized, mark the spot on the thermostat dial. It may also be a good idea to tape the dial in position so that it is not moved accidentally.

4. Place the eggs inside the incubator with the *X* mark on each egg in the up position.

5. Place a shallow pan of water in the incubator. As the water evaporates, it will keep the air humid. Check the humidity each day. Refill the pan as necessary.

6. Keep a record of the temperature and humidity. Record this information in Table 3.5.2 in the Results and Discussion section.

7. The eggs must be turned at least twice per day up to day eighteen. After this time the eggs are not turned because the chick is getting into position to hatch. An incubator that turns eggs mechanically is recommended. If you must turn eggs by hand, turn all the *X*s up, then on the next turn, place all the blank sides up.

8. Examine the eggs each day and discard leaking, cracked, or moldy eggs. These eggs may be infected with microbes. Eggs with a high microbe population build up pressure and may explode causing damage to the developing eggs - not to mention the mess!

9. Candling may be done each day up to day eighteen if great care is taken. Follow your teacher's instruction or instructions that came with the candler. This will help you in completing the Results and Discussion section. Remove any clears (infertile eggs) and eggs with dead embryos.

10. A hatching rate of 60%-80% is considered good. Chicks will begin to hatch at the end of the twentieth day. Healthy chicks can emerge from the shell without assistance. Those that have difficulty probably will not survive even if you help. Difficulty in emerging usually reflects a problem in incubation, such as low humidity or improper rotation.

11. After the chicks hatch, they can be left in the incubator for a few hours but should soon be transferred to a brooder with a temperature of about 95°F. Reduce the temperature 5° each week until room temperature is reached.

RESULTS AND DISCUSSION

1. What do you think will happen to the weight of the egg during incubation? Will it increase, decrease, or remain the same?

2. Weigh your egg before putting it into the incubator and each day up to day eighteen. Use Table 3.5.1 to record egg weights.

Table 3.5.1 Weight Change in Incubating Eggs

Day	Egg Weight	Difference
1		
2		
3		
4		
5		
6		
7		
8		
9		
10		
11		
12		
13		
14		
15		
16		
17		

3. Prepare a graph in the space below to show the change in egg weight. Use the data you collected.

4. How do you explain the change in weight of the egg?

5. Record the temperature and humidity in the incubator each day in Table 3.5.2.

Table 3.5.2 Record of Humidity and Temperature for Hatching Chicks

Day	Room Temperature	Dry Bulb Temperature	Wet Bulb Temperature	Humidity
1				
2				
3				
4				
5				
6				
7				
8				
9				
10				
11				
12				
13				
14				
15				
16				
17				
18				
19				
20				
21				

6. Why must the chick hatch on the twenty-first day?

NAME:______________________ DATE:____________ CLASS:____________

Chapter 3: The Poultry Industry

Laboratory Exercise 3.6 The Value of Poultry Products

BACKGROUND

Poultry products are currently very popular in the United States. Chicken, turkey, and eggs account for 8.7% of farm income for all products. The consumption of all poultry meat has nearly doubled in the past twenty years. The consumption of turkey more than doubled over the last twenty years, although Americans still eat nearly four times as much chicken as they do turkey.

In this exercise you will use statistics from the USDA to determine the top producing states for chicken, turkey, and eggs. You will also review your state's position in the production of poultry products relative to other states.

OBJECTIVES

- To compare the production of poultry products among the states
- To compare the production of poultry products in your state to other states
- To identify on a map the states and regions highest in poultry production in the United States
- To determine the per capita consumption of chicken, turkey, and eggs

---

EQUIPMENT

No special equipment is needed for this exercise.

MATERIALS

References:

Ranking of States and Commodities by Cash Receipts (latest edition). Economics Research Service, United States Department of Agriculture, Herndon, VA 22070. Telephone 1-800-999-6779.

Statistical Abstract of the United States (latest edition). U.S. Department of Commerce, Bureau of the Census, Washington, DC 20402-9328.

SAFETY

No special safety precautions are necessary with this exercise.

PROCEDURES

1. Use the references to answer the items in the Results and Discussion section. Your teacher may direct you to use Tables 3.6.2 through 3.6.7 at the end of this exercise instead of the references listed.

RESULTS AND DISCUSSION

1. What is the per capita consumption of chicken, turkey, and eggs in the United States? Complete Table 3.6.1 using the data found in the Statistical Abstract of the United States, Per Capita Consumption of Major Food Commodities. Draw arrows, up or down (↑ or ↓), to indicate if consumption has gone up or down from the first year reported to the last year reported.

Table 3.6.1 Per Capita Consumption of Chicken, Turkey, and Eggs

Commodity	Consumption Per Capita		
	19___	19___	Trend ↑ or ↓
Poultry Products (pounds)			
Chicken (pounds)			
Turkey (pounds)			
Eggs (number)			

2. How does the per capita consumption of poultry compare with what you and your family eat?

Do you and your family eat more or less

Chicken? ______________________________

Turkey? ______________________________

Eggs? ______________________________

What are the possible reasons for these differences and/or similarities?

3. List the five leading states in the production of poultry and eggs.

1.

2.

3.

4.

5.

4. The top five states produce _____ % of poultry and eggs in the United States.

5. Compile the following statistics for your state:

My state ranks ______________ in the production of poultry and eggs

The dollar value of poultry and eggs in my state is ______________ .

My state produces ________ % of the total value of poultry and eggs produced in the United States.

The sale of poultry and eggs represents ________ % of the value of agricultural commodities in my state.

6. List the five leading states in the production of broilers.

   1.

   2.

   3.

   4.

   5.

7. The top five states produce ______ % of the value of broilers in the United States.

8. Compile the following statistics for your state:

   My state ranks ___________ in the production of broilers.

   The dollar value of broilers in my state is ____________ .

   My state produces ________ % of the total value of broilers produced in the United States.

   The sale of broilers represents __________ % of the value of agricultural commodities in my state.

9. List the five leading states in the production of turkeys.

   1.

   2.

   3.

   4.

   5.

10. The top five states produce ______ % of the value of all turkeys produced in the United States.

11. Compile the following statistics for your state:

My state ranks ______________________ in the production of turkeys.

The dollar value of turkeys in my state is ______________________ .

My state produces ________ % of the total value of turkeys produced in the United States.

The sale of turkeys represents __________ % of the value of agricultural commodities in my state.

12. List the five leading states in the production of eggs.

1.

2.

3.

4.

5.

13. The top five states produce ______ % of the value of eggs produced in the United States.

14. Compile the following statistics for your state:

My state ranks ________________________ in the production of eggs.

The dollar value of eggs in my state is ________________________ .

My state produces ________ % of the total value of eggs produced in the United States.

The sale of eggs represents ______ % of the value of agricultural commodities in my state.

15. Identify the top ten states in the production of broilers, turkeys, and eggs by marking the map in Figure 3.6.1. Use the following code to identify the top ten states in the production of each. Write the first letter of broiler, turkey, or egg and the rank of the state in production.

    For example:

    B-6 means that the state ranks 6th in the production of broilers.

    T-2 means that the state ranks 2d in turkey production.

    E-3 means that the state ranks 3d in egg production.

16. List and briefly discuss the factors that affect the numbers of broilers, turkeys, and eggs produced in your state.

17. If a state does not rank in the top ten in the production of poultry products, does that mean that poultry products are relatively unimportant to that state's economy? Explain your answer.

18. How do broilers rank in value in the United States as compared to other commodities? ________________

19. How do turkeys rank in value in the United States as compared to other commodities? ________________

20. How do eggs rank in value in the United States as compared to other commodities? ________________

21. Broilers, turkeys, and eggs together represent __________% of the total United States receipts for farm commodities.

Table 3.6.2 Per Capita Consumption of Selected Food Commodities (pounds): 1970 to 1990

Commodity	1970	1975	1980	1985	1990
Red meat, total	132.0	125.3	126.4	124.9	112.3
Beef	79.6	83.0	72.1	74.6	64.0
Veal	2.0	2.8	1.3	1.5	.09
Lamb and Mutton	2.1	1.3	1.0	1.1	1.1
Pork	48.2	38.2	52.1	47.7	46.3
Fish and Shellfish	11.8	12.2	12.5	15.1	15.5
Poultry products, total	34.1	34.2	42.6	49.4	63.6
Chicken	27.7	27.5	34.3	39.9	49.3
Turkey	6.4	6.7	8.3	9.6	14.4
Eggs (number)	309	276	271	255	233
Milk, total (milk equivalent)	563.8	539.1	543.3	593.7	570.6
Fluid milk and cream	275.1	261.4	245.6	241.0	233.2
Milk, beverages	269.1	254.0	237.4	229.7	221.5
Cheese	11.4	14.3	17.5	22.5	24.7
Ice cream	17.8	18.6	17.5	18.1	15.7

Table 3.6.3 Poultry and Eggs: States' Rankings for Cash Receipts, 1991

State and Rank		Value of commodity group receipts	Percent of commodity group total	Cumulative percent 1/	Percent of State's total for all commodities	State's total for all commodities
		1,000 dollars	..........Percent..........			1,000 dollars
1	Arkansas	1,852,072	12.3	12.3	42.9	4,310,724
2	North Carolina	1,516,882	10.0	22.3	30.8	4,924,071
3	Georgia	1.446,931	9.6	31.9	36.3	3,978,361
4	Alabama	1,317,776	8.7	40.7	44.2	2,977,832
5	California	964,376	6.4	47.1	5.3	17,886,698
6	Texas	787,903	5.2	52.3	6.5	12,126,182
7	Mississippi	689,057	4.5	56.9	28.4	2,422,070
8	Virginia	516,917	3.4	60.3	24.6	2,095,371
9	Minnesota	506,154	3.3	63.7	7.3	6,936,001
10	Indiana	481,180	3.1	66.9	10.7	4,474,513
11	Pennsylvania	470,351	3.1	70.0	13.4	3,503,040
12	Maryland	444,081	2.9	72.9	33.3	1,332,494
13	Missouri	431,566	2.8	75.8	11.1	3,861,179
14	Delaware	407,995	2.7	78.5	65.8	619,536
15	Ohio	311,321	2.0	80.6	8.0	3,893,074
16	Oklahoma	279,390	1.8	82.4	7.3	3,807,582
17	Florida	277,068	1.8	84.3	4.5	6,140,999
18	South Carolina	270,229	1.7	86.1	22.0	1,225,396
19	Iowa	224,362	1.4	87.6	2.2	10,179,249
20	Louisiana	205,438	1.3	88.9	11.4	1,792,907
21	Wisconsin	172,593	1.1	90.1	3.1	5,449,043
22	Tennessee	159,782	1.0	91.1	8.0	1,977,569
23	Washington	134,608	0.8	92.0	3.4	3,946,524
24	Colorado	125,267	0.8	92.9	3.3	3,761,320
25	Michigan	113,545	0.7	93.6	3.6.	3,081,072
26	Oregon	102,125	0.6	94.3	4.1	2,454,389
27	West Virginia	100,261	0.6	95.0	30.3	330,237
28	Maine	90,295	0.6	95.5	20.3	444,601
29	Connecticut	89,229	0.5	96.1	19.2	463,372
30	Nebraska	77,512	0.5	96.7	0.8	8,821,328
31	New York	74,467	0.4	97.2	2.6	2,868,321
32	Utah	69,419	0.4	97.6	9.5	730,882
33	Illinois	68,973	0.4	98.1	0.9	7,508,777
34	South Dakota	56,948	0.3	98.4	1.7	3,264,286
35	Kentucky	54,720	0.3	98.8	1.7	3,178,704
36	New Jersey	37,555	0.2	99.1	5.6	660,160
37	Massachusetts	23,662	0.1	99.2	4.9	475,540
38	Kansas	20,940	0.1	99.4	0.3	6,934,986
39	Hawaii	19,392	0.1	99.5	3.2	596,925
40	New Mexico	17,699	0.1	99.6	1.1	1,501,152
41	Idaho	12,443	0.0	99.7	0.4	2,615,946
42	North Dakota	11,255	0.0	99.8	0.4	2,556,147
43	Montana	9,161	0.0	99.8	0.6	1,531,169
44	New Hampshire	8,079	0.0	99.9	5.6	143,106
45	Rhode Island	4,819	0.0	99.9	6.8	70,917
46	Arizona	3,996	0.0	99.9	0.2	1,889,907
47	Vermont	2,924	0.0	100.0	0.6	433,140
48	Wyoming	259	0.0	100.0	0.0	812,743
49	Alaska	86	0.0	100.0	0.3	26,622
50	Nevada	80	0.0	100.0	0.0	275,836
	United States	15,063,143			9.0	167,292.000

Numbers may not add due to rounding. 1/ The cumulative percentage is the sum of the percent of commodity group total for each State and all preceding States.

Table 3.6.4 Broilers: States' Rankings for Cash Receipts, 1991

State and Rank		Value of commodity receipts	Percent of commodity total	Cumulative percent 1/	Percent of State's total for all commodities	State's total for all commodities
		1,000 dollars		....Percent....		1,000 dollars
1	Arkansas	1,369,830	16.3	16.3	31.7	4,310,724
2	Alabama	1,147,956	13.6	30.0	38.5	2,977,832
3	Georgia	1,125,755	13.4	43.4	28.3	3,978,361
4	North Carolina	777,003	9.2	52.7	15.7	4,924,071
5	Mississippi	579,070	6.9	59.6	23.9	2,422,070
6	Texas	508,939	6.0	65.6	4.2	12,126,182
7	Delaware	391,999	4.6	70.3	63.2	619,536
8	Maryland	384,384	4.5	74.9	28.8	1,332,494
9	California	334,080	3.9	78.9	1.8	17,886,698
10	Virginia	305,087	3.6	82.5	14.5	2,095,371
11	Oklahoma	189,453	2.2	84.8	4.9	3,807,582
12	Pennsylvania	174,013	2.0	86.9	4.9	3,503,040
13	Missouri	155,610	1.8	88.7	4.0	3,861,179
14	Florida	151,704	1.8	90.5	2.4	6,140,999
15	Tennessee	140,444	1.6	92.2	7.1	1,977,569
16	South Carolina	116,475	1.3	93.6	9.5	1,225,396
17	Minnesota	70,950	0.8	94.4	1.0	6,936,001
18	West Virginia	54,895	0.6	95.1	16.6	330,237
19	Washington	53,976	0.6	95.7	1.3	3,946,524
20	Oregon	34,253	0.4	96.1	1.4	2,454,389
21	Ohio	30,263	0.3	96.5	0.7	3,893,074
22	Iowa	27,495	0.3	96.8	0.2	10,179,249
23	Kentucky	24,864	0.3	97.1	0.7	3,178,704
24	Wisconsin	19,382	0.2	97.4	0.3	5,449,043
25	Nebraska	5,544	0.0	97.4	0.0	8,821,328
26	Hawaii	3,214	0.0	97.5	0.5	596,925
27	New York	2,835	0.0	97.5	0.1	2,868,321
28	Michigan	702	0.0	97.5	0.0	3,081,072
	Louisiana	2/				
	Indiana	2/				
	South Dakota	2/				
	North Dakota	2/				
	United States	8,385,284			5.0	167,292,000

Numbers may not add due to rounding.

1/ The cumulative percentage is the sum of the percent of commodity total for each State and all preceding States.

2/ States at the bottom of the above ranked list of States and having no accompanying data would have appeared within the list of 50 states' rankings, but were excluded to avoid disclosure of confidential information about individual producers.

Table 3.6.5 Turkeys: States' Rankings for Cash Receipts, 1991

State and Rank		Value of commodity receipts	Percent of commodity total	Cumulative percent 1/	Percent of State's total for all commodities	State's total for all commodities
		1,000 dollars	..............Percent..............			1,000 dollars
1	North Carolina	431,346	18.4	18.4	8.7	4,924,071
2	Minnesota	289,872	12.3	30.7	4.1	6,936,001
3	California	241,425	10.3	41.0	1.3	17,886,698
4	Arkansas	186,048	7.9	49.0	4.3	4,310,724
5	Missouri	171,570	7.3	56.3	4.4	3,861,179
6	Indiana	134,190	5.7	62.0	3.0	4,474,513
7	Virginia	122,934	5.2	67.2	5.8	2,095,371
8	Iowa	90,480	3.8	71.1	0.8	10,179,249
9	Pennsylvania	68,880	2.9	74.0	1.9	3,503,040
10	South Carolina	65,934	2.8	76.9	5.3	1,225,396
11	Ohio	50,580	2.1	79.0	1.3	3,893,074
12	Michigan	49,350	2.1	81.1	1.6	3,081,072
13	Utah	45,158	1.9	83.1	6.1	730,882
14	South Dakota	29,160	1.2	84.3	0.8	3,264,286
15	West Virginia	29,135	1.2	85.5	8.8	330,237
16	Illinois	26,994	1.1	86.7	0.3	7,508,777
17	Georgia	20,879	0.8	87.6	0.5	3,978,361
18	Oregon	14,805	0.6	88.2	0.6	2,454,389
19	Nebraska	13,841	0.5	88.8	0.1	8,821,328
20	North Dakota	8,564	0.3	89.2	0.3	2,556,147
21	Kansas	4,894	0.2	89.4	0.0	6,934,986
22	New York	4,387	0.1	89.6	0.1	2,868,321
23	Massachusetts	3,230	0.1	89.7	0.6	475,540
24	New Jersey	1,393	0.0	89.8	0.2	660,160
25	New Hampshire	622	0.0	89.8	0.4	143,106
26	Connecticut	393	0.0	89.8	0.0	463,372
	Wisconsin	2/				
	Texas	2/				
	Colorado	2/				
	Oklahoma	2/				
	Maryland	2/				
	Delaware	2/				
	United States	2,344,016			1.4	167,292,000

Numbers may not add due to rounding.

1/ The cumulative percentage is the sum of the percent of commodity total for each State and all preceding States.

2/ States at the bottom of the above ranked list of States and having no accompanying data would have appeared within the list of 50 states' rankings, but were excluded to avoid disclosure of confidential information about individual producers.

Table 3.6.6 Chicken Eggs: States' Rankings for Cash Receipts, 1991

State and Rank		Value of commodity receipts	Percent of commodity total	Cumulative percent 1/	Percent of State's total for all commodities	State's total for all commodities
		1,000 dollars	..........................Percent..........................			1,000 dollars
1	California	362,275	9.3	9.3	2.0	17,886,698
2	Georgia	289,601	7.5	16.8	7.2	3,978,361
3	Arkansas	285,881	7.4	24.2	6.6	4,310,724
4	Indiana	264,941	6.8	31.1	5.9	4,474,513
5	Pennsylvania	222,300	5.7	36.9	6.3	3,503,040
6	Ohio	211,370	5.4	42.3	5.4	3,893,074
7	Texas	203,038	5.2	47.6	1.6	12,126,182
8	North Carolina	199,701	5.1	52.8	4.0	4,924,071
9	Alabama	166,500	4.3	57.1	5.5	2,977,832
10	Florida	120,930	3.1	60.2	1.9	6,140,999
11	Mississippi	106,063	2.7	63.0	4.3	2,422,070
12	Minnesota	102,486	2.6	65.6	1.4	6,936,001
13	Iowa	99,243	2.5	68.2	0.9	10,179,249
14	Maine	86,492	2.2	70.4	19.4	444,601
15	Connecticut	82,160	2.1	72.5	17.7	463,372
16	Washington	78,233	2.0	74.6	1.9	3,946,524
17	South Carolina	77,982	2.0	76.6	6.3	1,225,396
18	Missouri	69,611	1.8	78.4	1.8	3,861,179
19	Oklahoma	69,167	1.7	80.2	1.8	3,807,582
20	Virginia	66,855	1.7	81.9	3.1	2,095,371
21	Michigan	62,587	1.6	83.5	2.0	3,081,072
22	Nebraska	57,050	1.4	85.0	0.6	8,821,328
23	Maryland	56,424	1.4	86.5	4.2	1,332,494
24	Colorado	53,108	1.3	87.9	1.4	3,761,320
25	New York	52,640	1.3	89.2	1.8	2,868,321
26	Oregon	42,646	1.1	90.3	1.7	2,454,389
27	Illinois	40,989	1.0	91.4	0.5	7,508,777
28	Wisconsin	39,285	1.0	92.4	0.7	5,449,043
29	New Jersey	33,961	O.8	93.3	5.1	660,160
30	Kentucky	29,584	0.7	94.0	0.9	3,178,704
31	Louisiana	24,342	0.6	94.7	1.3	1,792,907
32	Utah	23,895	0.6	95.3	3.2	730,882
33	South Dakota	22,976	0.5	95.9	0.7	3,264,286
34	Massachusetts	20,145	0.5	96.4	4.2	475,540
35	Tennessee	18,763	0.4	96.9	0.9	1,977,569
36	New Mexico	17,617	0.4	97.4	1.1	1,501,152
37	Hawaii	15,996	0.4	97.8	2.6	596,925
38	West Virginia	15,805	0.4	98.2	4.7	330,237
39	Kansas	15,495	0.4	98.6	0.2	6,934,986
40	Delaware	15,443	0.4	99.0	2.4	619,536
41	Idaho	12,332	0.3	99.3	0.4	2,615,946
42	Montana	9,020	0.2	99.5	0.5	1,531,169
43	New Hampshire	4,247	0.1	99.6	2.9	143,106
44	Rhode Island	4,125	0.1	99.7	5.8	70,917
45	Arizona	3,453	0.0	99.8	0.1	1,889,907
46	Vermont	2,550	0.0	99.9	0.5	433,140
47	North Dakota	1,800	0.0	99.9	0.0	2,556,147
48	Wyoming	118	0.0	100.0	0.0	812,743
49	Nevada	79	0.0	100.0	0.0	275,836
50	Alaska	54	0.0	100.0	0.2	26,622
	United States	3,861,358			2.3	167,292,000

Numbers may not add due to rounding. 1/ The cumulative percentage is the sum of the percent of commodity total for each State and all preceding States.

Table 3.6.7 United States: Leading Commodities for Cash Receipts, 1991

Item and Rank		Value of U.S. receipts	Percent of U.S. total	Cumulative percent 1/	Rank in prior year
		1,000 dollars	................Percent...........		
	All commodities	167,292,000	100.0		
	livestock and products	86,745,278	51.8		
	Crops	80,546,722	48.1		
1	Cattle and calves	39,632,088	23.6	23.6	1
2	Dairy products	18,113,714	10.8	34.5	2
3	Corn	13,853,798	8.2	42.8	3
4	Hogs	11,061,441	6.6	49.4	4
5	Soybeans	10,778,421	6.4	55.8	5
6	Greenhouse and nursery	8,404,722	5.0	60.8	7
7	Broilers	8,385,284	5.0	65.8	6
8	Wheat	5,715,687	3.4	69.3	8
9	Cotton	5,588,934	3.3	72.6	9
10	Chicken eggs	3,861,358	2.3	74.9	10
11	Hay	3,044,049	1,8	76.7	11
12	Tobacco	2,886,039	1.7	78.5	12
13	Turkeys	2,344,016	1.4	79.9	14
14	Potatoes	2,047,785	1.2	81.1	13
15	Tomatoes	1,798.806	1.0	82.2	17
16	Apples	1,659,334	0.9	83.1	20
17	Grapes	1,618,558	0.9	84.1	16
18	Oranges	1,564,762	0.9	85.1	15
19	Peanuts	1,392,537	0.8	85.9	18
20	Sugar beets	1,216,758	0.7	86.6	19
21	Sorghum grain	1,131,780	0.6	87.3	22
22	Rice	1,092,385	0.6	87.9	21
23	Cane for sugar	902,031	0.5	88.5	24
24	Barley	821,519	0.4	89.0	25
25	Lettuce	817,667	0.4	89.5	23
Government payments 2/		8,214,399			

Numbers may not add due to rounding. 1/ The cumulative percentage is the sum of the percent of U.S. total for each commodity and all preceding commodities.
2/ Government payment made directly to farmers in cash or Payment-in-Kind.

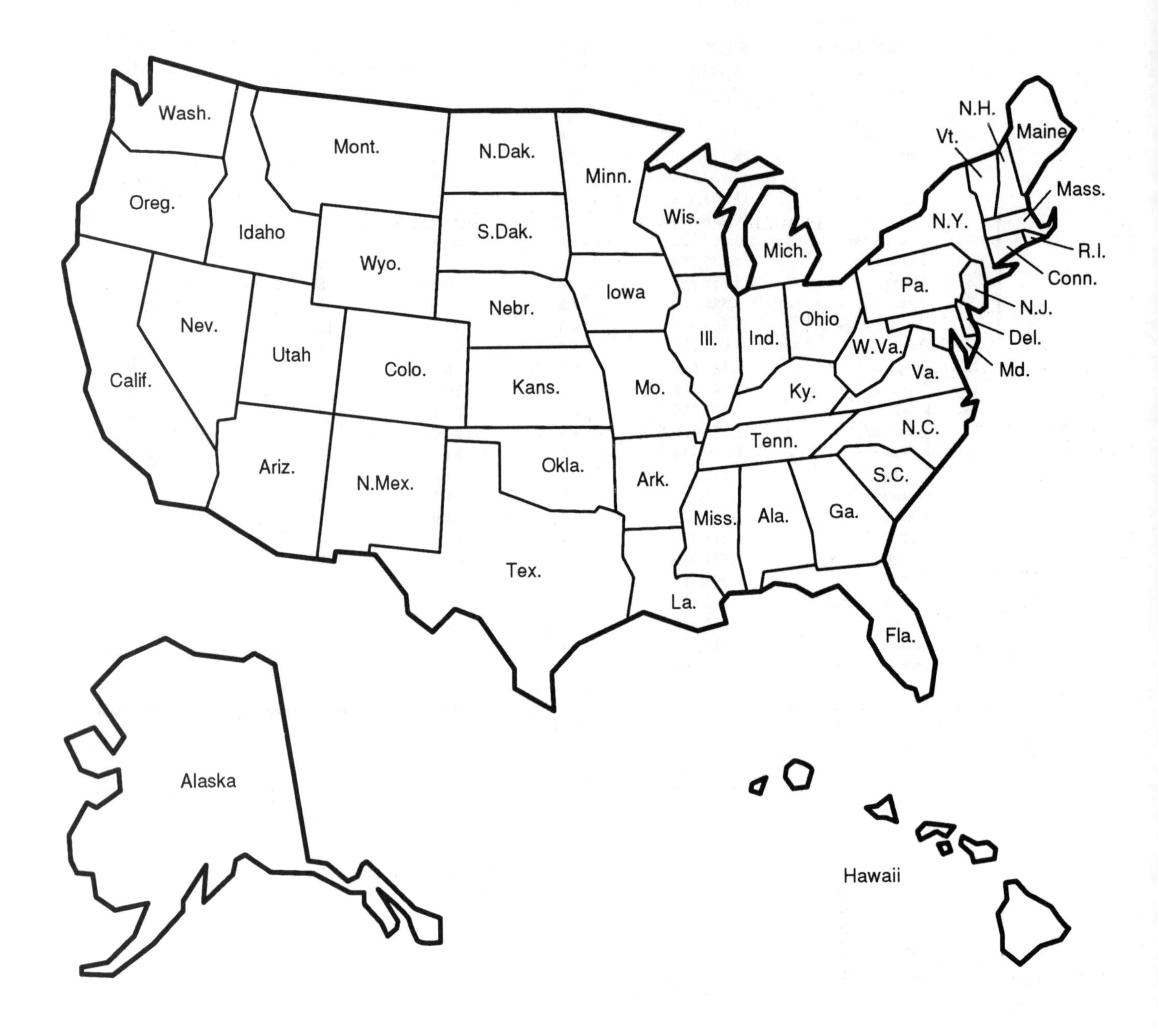
Wash.
Mont.
N.Dak.
Minn.
Vt.
N.H.
Maine
Oreg.
Idaho
S.Dak.
Wis.
Mass.
N.Y.
Wyo.
Mich.
R.I.
Conn.
Iowa
Pa.
Nebr.
N.J.
Nev.
Ill.
Ind.
Ohio
Del.
Utah
W.Va.
Md.
Colo.
Va.
Calif.
Kans.
Mo.
Ky.
N.C.
Tenn.
Okla.
Ariz.
N.Mex.
Ark.
S.C.
Miss.
Ala.
Ga.
Tex.
La.
Fla.
Alaska
Hawaii

Figure 3.6.1

NAME:______________________ DATE:____________ CLASS:____________

Chapter 4: The Dairy Industry

Laboratory Exercise 4.1 Making Ice Cream

BACKGROUND

Ice cream is a favorite dessert for many people. It was considered a special treat back in the 1700s and 1800s because not everyone could purchase the equipment to make homemade ice cream. In 1851 Jacob Fussel started the first wholesale ice cream business.

Today the United States consumes approximately one billion gallons of ice cream and ice cream related products per year. Many people worry about the calories in ice cream due to the high fat content and the amount of sugar used in making ice cream. However, ice cream is also nutritious. It contains proteins, vitamins, and calcium.

The physical structure of ice cream is a foam that has air cells and forms overrun. Overrun is essential because it gives the ice cream twice the original volume and a light, smooth texture. Ice cream is composed mainly of milk solids and milk fat, sugar, and stabilizing/emulsifying ingredients. The milk fat gives flavor and improves the body and texture of ice cream. Milk solids that are not fat also contribute to the flavor and texture. Sugar adds sweetness to the ice cream and helps lower the freezing temperature and prevents the ice cream from freezing solid in the freezer. The stabilizers aid in reaching and maintaining the desired product and help improve the texture. Emulsifiers aid in scattering the fat globules throughout the ice cream and improve its whipping characteristics.

In this laboratory exercise you will make several different types of homemade ice cream and compare their textures and flavors. You will also learn about the different dairy products that are used in making ice cream.

OBJECTIVES

- To identify the ingredients and procedures necessary for making ice cream
- To explain the importance of ice cream products to the agriculture industry
- To describe any differences in taste and texture of ice creams made from various recipes

---

EQUIPMENT

one ice cream freezer - electric or hand crank
large mixing bowls for mixing ingredients
large cooking pots, if needed per recipe instructions
bowls and spoons for taste tests

Materials

ingredients necessary for each recipe used
rock salt
ice

Safety

Follow manufacturer's directions or your teacher's instructions on how to operate the ice cream freezers.
Do not use recipes that include raw eggs.

PROCEDURES

1. Thoroughly wash and sanitize can, paddle, and lid of ice cream freezer and any other utensil used in preparing the ice cream.

   NOTE: Equipment can be sanitized by immersing the parts in 170° water for one minute.

2. Assemble ice cream freezer following manufacturer's instructions or directions given by your teacher.

3. You will be making two batches of ice cream and one batch of sherbet. Mix the ingredients according to recipe directions, except in one batch of ice cream substitute Half-and-Half for the cream. Pour the mixture into the sanitized can in ice cream freezer.

4. Place lid on can.

5. Place top on ice cream freezer following manufacturer's or instructor's directions.

6. Plug in freezer or if using a hand crank, start turning handle.

7. Add ice and salt in alternate layers until tub is full but **not** over the top of the can. As ice melts, continue adding ice and salt to keep the level of ice at the top of the can but **not** over the top.

8. Make sure the drain hole remains unclogged during freezing.

9. Continue adding ice and salt until the motor stops or turning by hand becomes too difficult.

The following recipes may be used for this lab activity, or you may bring your favorite recipe from home.

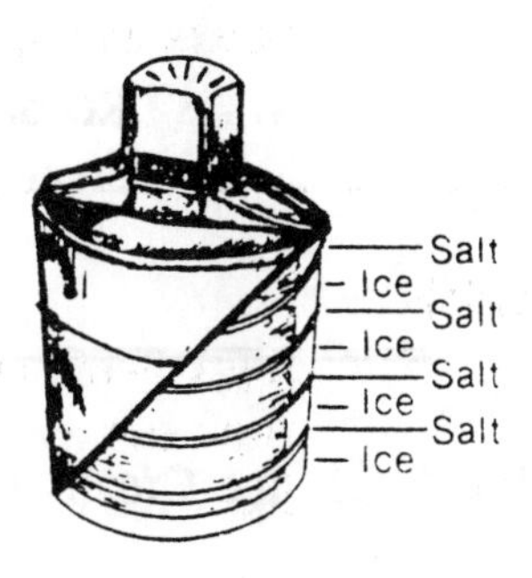

Figure 4.1.1 Ice and salt should be layered in the churn

Vanilla Ice Cream

2 cups milk
1 cup sugar
2 tablespoons vanilla
1/4 teaspoon salt
4 cups light cream

Heat milk until it just starts to boil, remove from heat. Add the sugar and salt to the milk and stir until both are well dissolved. Add the cream. Stir in the vanilla. Cool. Pour mixture into ice cream freezer and freeze following the above instructions. Makes approximately two quarts.

Fruit Sherbet

4 cups milk
2 tablespoons lemon juice
1/2 cup mild-flavored honey
2 1/2 cups fruit juice and/or puree or pulp

In a medium-size bowl thoroughly combine milk, honey, lemon juice, and fruit juice and/or puree or pulp. Some curdling may occur, but this will not affect the sherbet. Pour into an ice cream maker and process. Makes approximately two quarts.

RESULTS AND DISCUSSION

1. What dairy products are used in making ice cream?

2. Why is it important to sanitize the equipment used to make ice cream?

3. Why are dairy products an important part of the agricultural industry?

4. Record your description of the flavor and texture of the different ice creams and sherbet in Table 4.1.1.

Table 4.1.1 Ice Cream Taste Test

	Flavor	Texture	Other Comments
Ice cream made with light cream			
Ice cream made with Half-and-Half			
Sherbet			

5. Ice cream made using the recipe in this lab with light cream contains about 12 to 13% milk fat. When Half-and-Half is substituted for the cream, the fat content is about 7 to 8% (dairy portion). How does fat content affect the flavor and texture of ice cream?

6. Which of the ice cream products tasted most like ice cream and which tasted most like ice milk?

NAME:_______________ DATE:_______________ CLASS:_______________

Chapter 4: The Dairy Industry

Laboratory Exercise 4.2 Making Butter

BACKGROUND

The dairy industry is an important component of agricultural production in the United States. An average family of four in the United States consumes approximately 1,233 pounds of dairy products per year. The dairy industry encompasses beverage milk, cream, cheese, butter, ice cream, etc.

Products made from milk and cream have become a part of most Americans' daily diet. It has only been in the last fifty to sixty years that butter has become a regular part of meals. Back in the 1930s, the cream would be collected from the top of the milk and placed in a wooden butter churn. It would be churned by hand in most homes and would be used only on special occasions, such as holidays and when company would come for a visit.

The USDA defines butter as consisting of not less than 80% milk fat and not more than 16% water. Butter can be made by using the cream that has separated from raw milk or by purchasing whipping cream and churning by hand or mixing in a blender. Once the butter has formed, the butter will have to be separated from the buttermilk. Once the butter is made, it can then be molded and placed in the refrigerator to harden, or it may be eaten immediately.

In this laboratory exercise you will make a small amount of butter by shaking whipping cream in a jar. This simulates the processing of butter in a churn. A churn is a simple machine that makes the process much easier, especially in larger quantities.

OBJECTIVES

- To explain the importance of the dairy industry
- To identify the steps in making butter
- To make butter using whipping cream

EQUIPMENT

one plastic soda bottle with cap
one plastic knife or fork

toaster (optional)
two bowls or cups (one for butter and one for the buttermilk)
scissors
laboratory scales

MATERIALS

one-half pint whipping cream
bread (for making toast, optional) or crackers

SAFETY

Some people are allergic to dairy products. This exercise directs you to taste the butter that is made. Be sure to tell your teacher if you have allergies to dairy products and do not taste any of the products.

PROCEDURES

1. Obtain a soda bottle with a cap and wash both items in warm, soapy water.

2. Pour one-half pint of whipping cream into the bottle. Place the lid on the container and tighten.

3. Shake the container vigorously until butter forms - about twenty minutes.

4. You will begin to notice a change in the consistency of the cream in about ten minutes. Continue to shake until there is a distinct difference in the watery and solid substance in the container.

5. Once butter has formed, separate the butter from the buttermilk. Pour the buttermilk from the bottle into a cup.

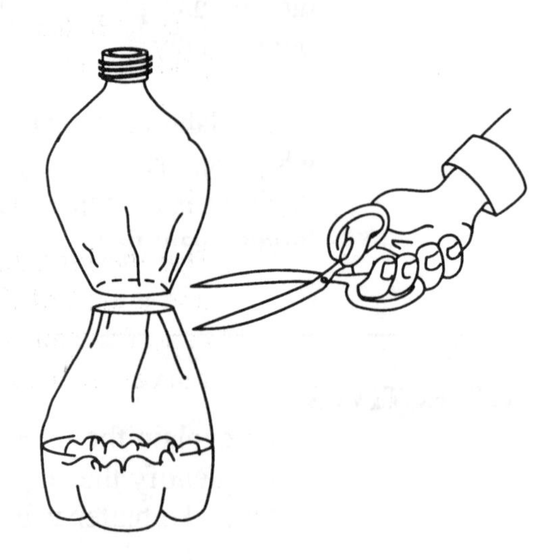

Figure 4.2.1 Squeeze the soda bottle in half and cut off the top to remove the butter

6. Then remove the butter. Most plastic soda bottles will have to be cut open to remove the butter. Do this very carefully and under the supervision of your teacher. Remove the bottle cap and squeeze the sides of the bottle together, as shown in Figure 4.2.1. Use scissors to cut the bottle in half.

7. Put the butter into a bowl. Weigh the butter and complete items 1 through 3 in the Results and Discussion section.

8. The butter is now ready to sample. Spread the butter onto a toasted slice of bread or crackers and taste.

9. You may want to mold the butter and set it in the refrigerator to harden. Special butter molds can be purchased, or use your imagination to shape the butter.

RESULTS AND DISCUSSION

1. Weigh the butter you just made (be sure to subtract the weight of the bowl). Enter your answer in ounces:

2. Butter can be purchased at the grocery store for about $1.75 per pound. Ask your teacher the price of the cream you just used and enter the price here:

3. Use the information from #1 and #2 for this item. How much would one pound of the butter you made cost, based only on the price of the cream? You will need to divide the cost of the cream by the number of ounces of butter and multiply the cost by 16 (16 ounces per pound).

4. Which costs more, the butter you made or butter purchased from the store? Explain why this is the case.

5. Describe the flavor of the butter you just made.

6. After cooling the butter in the refrigerator, try spreading the butter on bread. What is the result? Why is this a problem in marketing butter?

7. List as many dairy products as you can that your family consumes and compare them with the dairy products consumed by other students and their families in your class. From the information collected fill in Table 4.1.1.

Table 4.1.1 Dairy Products Consumed

8. Which dairy product is consumed in the largest amount in your home? What is the estimated amount consumed in one year?

9. Which dairy product is used the least in your home?

10. Why is the dairy industry an important part of agriculture?

NAME:______________________ DATE:____________ CLASS:____________

Chapter 4: The Dairy Industry

Laboratory Exercise 4.3 The Value of The Dairy Industry

BACKGROUND

Dairy products are an important part of the American diet. Dairy products account for 10.8% of income for all agricultural products. The consumption of all milk products has increased slightly during the past twenty years. The consumption of cheeses has more than doubled.

Figure 4.3.1

In this exercise you will use statistics from the USDA to determine the top dairy producing states. You will also review your state's position in the production of dairy products relative to other states.

OBJECTIVES

- To compare the production of dairy products among the states
- To compare the production of dairy products in your state as compared to other states
- To identify on a map the states and regions highest in dairy production in the United States
- To determine the per capita consumption of dairy products

EQUIPMENT

No special equipment is needed for this exercise.

MATERIALS

References:

Ranking of States and Commodities by Cash Receipts (latest edition). Economics Research Service, United States Department of Agriculture, Herndon, VA 22070. Telephone 1-800-999-6779.

Statistical Abstract of the United States (latest edition). U.S. Department of Commerce, Bureau of the Census, Washington, DC 20402-9328.

SAFETY

No special safety precautions are necessary with this exercise.

PROCEDURES

1. Use the references to answer the items in the Results and Discussion section. Your teacher may direct you to use Tables 4.3.2 through 4.3.4 at the end of this exercise instead of the references listed.

RESULTS AND DISCUSSION

1. List the five leading states in the production of dairy products.

   1.

   2.

   3.

   4.

   5.

2. The top five states produce _____ % of dairy products in the United States.

3. Compile the following statistics for your state:

   My state ranks ________________ in the production of dairy products.

   The dollar value of dairy products in my state is ________________ .

   My state produces _______ % of the total value of dairy products produced in the United States.

   The sale of dairy products represents __________ % of the value of agricultural commodities in my state.

4. Identify the top ten states in the production of dairy products by marking the map in Figure 4.3.1. Use the following code to identify the top ten states: Write the first letter of "dairy" and the rank of the state in dairy products production.

   For example:

   D-5 means that the state ranks 5th in the production of dairy products.

5. List and briefly discuss the factors that affect the amount of dairy products produced in your state.

6. If a state does not rank in the top ten in the production of dairy products, does that mean that dairy products are relatively unimportant to that state's economy? Explain your answer.

7. How do dairy products rank in value in the United States as compared to other commodities?

8. What is the per capita consumption of milk and milk products in the United States? Complete Table 4.3.1 using the data found in the Statistical Abstract of the United States, Per Capita Consumption of Major Food Commodities. Draw arrows, up or down (↑ or ↓), to indicate if consumption has gone up or down from the first year reported to the last year reported.

Table 4.3.1 Per Capita Consumption of Milk and Milk Products			
Commodity	Consumption Per Capita		
	19___	19___	Trend ↑ or ↓
Milk, total (milk equivalent)			
Fluid milk and cream			
Milk beverages			
Cheese			
Ice cream			

9. How does the per capita consumption of dairy products compare with what you and your family eat? Do you and your family consume more or less

   Milk? ____________________

   Cheese? ____________________

   Ice Cream? ____________________

   What are the possible reasons for these differences and/or similarities?

Table 4.3.2 Per Capita Consumption of Selected Food Commodities: 1970 - 1990					
Commodity	**1970**	**1975**	**1980**	**1985**	**1990**
Red meat, total	**132.0**	**125.3**	**126.4**	**124.9**	**112.3**
Beef	**79.6**	**83.0**	**72.1**	**74.6**	**64.0**
Veal	**2.0**	**2.8**	**1.3**	**1.5**	**.09**
Lamb and Mutton	**2.1**	**1.3**	**1.0**	**1.1**	**1.1**
Pork	**48.2**	**38.2**	**52.1**	**47.7**	**46.3**
Fish and Shellfish	**11.8**	**12.2**	**12.5**	**15.1**	**15.5**
Poultry products, total	**34.1**	**34.2**	**42.6**	**49.4**	**63.6**
Chicken	**27.7**	**27.5**	**34.3**	**39.9**	**49.3**
Turkey	**6.4**	**6.7**	**8.3**	**9.6**	**14.4**
Eggs (number)	**309**	**276**	**271**	**255**	**233**
Milk, total (milk equivalent)	**563.8**	**539.1**	**543.3**	**593.7**	**570.6**
Fluid milk and cream	**275.1**	**261.4**	**245.6**	**241.0**	**233.2**
Milk, beverages	**269.1**	**254.0**	**237.4**	**229.7**	**221.5**
Cheese	**11.4**	**14.3**	**17.5**	**22.5**	**24.7**
Ice cream	**17.8**	**18.6**	**17.5**	**18.1**	**15.7**

Table 4.3.3 Dairy Products: States' Rankings for Cash Receipts, 1991

	State and Rank	Value of commodity receipts	Percent of commodity total	Cumulative percent 1/	Percent of State's total for all commodities	State's total for all commodities
		1,000 dollars	Percent			1,000 dollars
1	Wisconsin	2,850,126	15.7	15.7	52.3	5,449,043
2	California	2,454,538	13.5	29.2	13.7	17,886,698
3	New York	1,386,601	7.6	36.9	48.3	2,868,321
4	Pennsylvania	1,346,826	7.4	44.3	38.4	3,503,040
5	Minnesota	1,148,720	6.3	50.7	16.5	6,936,001
6	Texas	683,260	3.7	54.4	5.6	12,126,182
7	Michigan	643,140	3.5	58.0	20.8	3,081,072
8	Ohio	592,455	3.2	61.3	15.2	3,893,074
9	Washington	557,943	3.0	64.3	14.1	3,946,524
10	Iowa	482,664	2.6	67.0	4.7	10,179,249
11	Florida	372,947	2.0	69.1	6.0	6,140,999
12	Illinois	338,550	1.8	70.9	4.5	7,508,777
13	Missouri	333,795	1.8	72.8	8.6	3,861,179
14	Idaho	318,570	1.7	74.5	12.1	2,615,946
15	Vermont	310,536	1.7	76.3	71.6	433,140
16	Indiana	270,435	1.4	77.7	6.0	4,474,513
17	Virginia	269,325	1.4	79.2	12.8	2,095,371
18	Kentucky	263,750	1.4	80.7	8.3	3,178,704
19	Tennessee	262,263	1.4	82.1	13.2	1,977,569
20	New Mexico	220,994	1.2	83.4	14.7	1,501,152
21	Arizona	207,644	1.1	84.5	10.9	1,889,907
22	North Carolina	204,267	1.1	85.6	4.1	4,924,071
23	Oregon	203,610	1.1	86.8	8.3	2,454,389
24	South Dakota	200,860	1.1	87.9	6.1	3,264,286
25	Georgia	199,073	1.1	89.0	5.0	3,978,361
26	Maryland	181,740	1.0	90.0	13.6	1,332,494
27	Colorado	166,156	0.9	90.9	4.4	3,761,320
28	Oklahoma	152,400	0.8	91.7	4.0	3,807,582
29	Utah	148,580	0.8	92.5	20.3	730,882
30	Nebraska	144,420	0.8	93.3	1.6	8,821,328
31	Kansas	141,453	0.7	94.1	2.0	6,934,986
32	Louisiana	121,695	0.6	94.8	6.7	1,792,907
33	North Dakota	114,695	0.6	95.4	4.4	2,556,147
34	Arkansas	102,700	0.5	96.0	2.3	4,310,724
35	Mississippi	96,862	0.5	96.5	4.0	2,422,070
36	Maine	85,060	0.4	97.0	19.1	444,601
37	Connecticut	70,647	0.3	97.4	15.2	463,372
38	Alabama	69,795	0.3	97.8	2.3	2,977,832
39	Massachusetts	64,977	0.3	98.1	13.6	475,540
40	South Carolina	55,073	0.3	98.4	4.4	1,225,396
41	New Jersey	52,407	0.2	98.7	7.9	660,160
42	New Hampshire	43,269	0.2	99.0	30.2	143,106
43	Montana	39,954	0.2	99.2	2.6	1,531,169
44	Nevada	38,396	0.2	99.4	13.9	275,836
45	West Virginia	33,678	0.1	99.6	10.2	330,237
46	Hawaii	30,328	0.1	99.8	5.0	596,925
47	Delaware	16,770	0.0	99.8	2.7	619,536
48	Wyoming	11,856	0.0	99.9	1.4	812,743
49	Rhode Island	5,329	0.0	99.9	7.5	70,917
50	Alaska	2,582	0.0	100.0	9.7	26,622
	United States	18,113,714			10.8	167,292,000

Numbers may not add due to rounding. 1/ The cumulative percentage is the sum of the percent of commodity total for each State and all preceding States.

Table 4.3.4 United States: Leading Commodities for Cash Receipts, 1991

Item and Rank		Value of U.S. receipts	Percent of U.S. total	Cumulative percent 1/	Rank in prior year
		1,000 dollars	...............Percent...........		
	All commodities	167,292,000	100.0		
	Livestock and Products	86,745,278	51.8		
	Crops	80,546,722	48.1		
1	Cattle and calves	39,632,088	23.6	23.6	1
2	Dairy products	18,113,714	10.8	34.5	2
3	Corn	13,853,798	8.2	42.8	3
4	Hogs	11,061,441	6.6	49.4	4
5	Soybeans	10,778,421	6.4	55.8	5
6	Greenhouse and nursery	8,404,722	5.0	60.8	7
7	Broilers	8,385,284	5.0	65.8	6
8	Wheat	5,715,687	3.4	69.3	8
9	Cotton	5,588,934	3.3	72.6	9
10	Chicken eggs	3,861,358	2.3	74.9	10
11	Hay	3,044,049	1,8	76.7	11
12	Tobacco	2,886,039	1.7	78.5	12
13	Turkeys	2,344,016	1.4	79.9	14
14	Potatoes	2,047,785	1.2	81.1	13
15	Tomatoes	1,798.806	1.0	82.2	17
16	Apples	1,659,334	0.9	83.1	20
17	Grapes	1,618,558	0.9	84.1	16
18	Oranges	1,564,762	0.9	85.1	15
19	Peanuts	1,392,537	0.8	85.9	18
20	Sugar beets	1,216,758	0.7	86.6	19
21	Sorghum grain	1,131,780	0.6	87.3	22
22	Rice	1,092,385	0.6	87.9	21
23	Cane for sugar	902,031	0.5	88.5	24
24	Barley	821,519	0.4	89.0	25
25	Lettuce	817,667	0.4	89.5	23
Government payments 2/		8,214,399			

Numbers may not add due to rounding. 1/ The cumulative percentage is the sum of the percent of U.S. total for each commodity and all preceding commodities.
2/ Government payment made directly to farmers in cash or Payment-in-Kind.

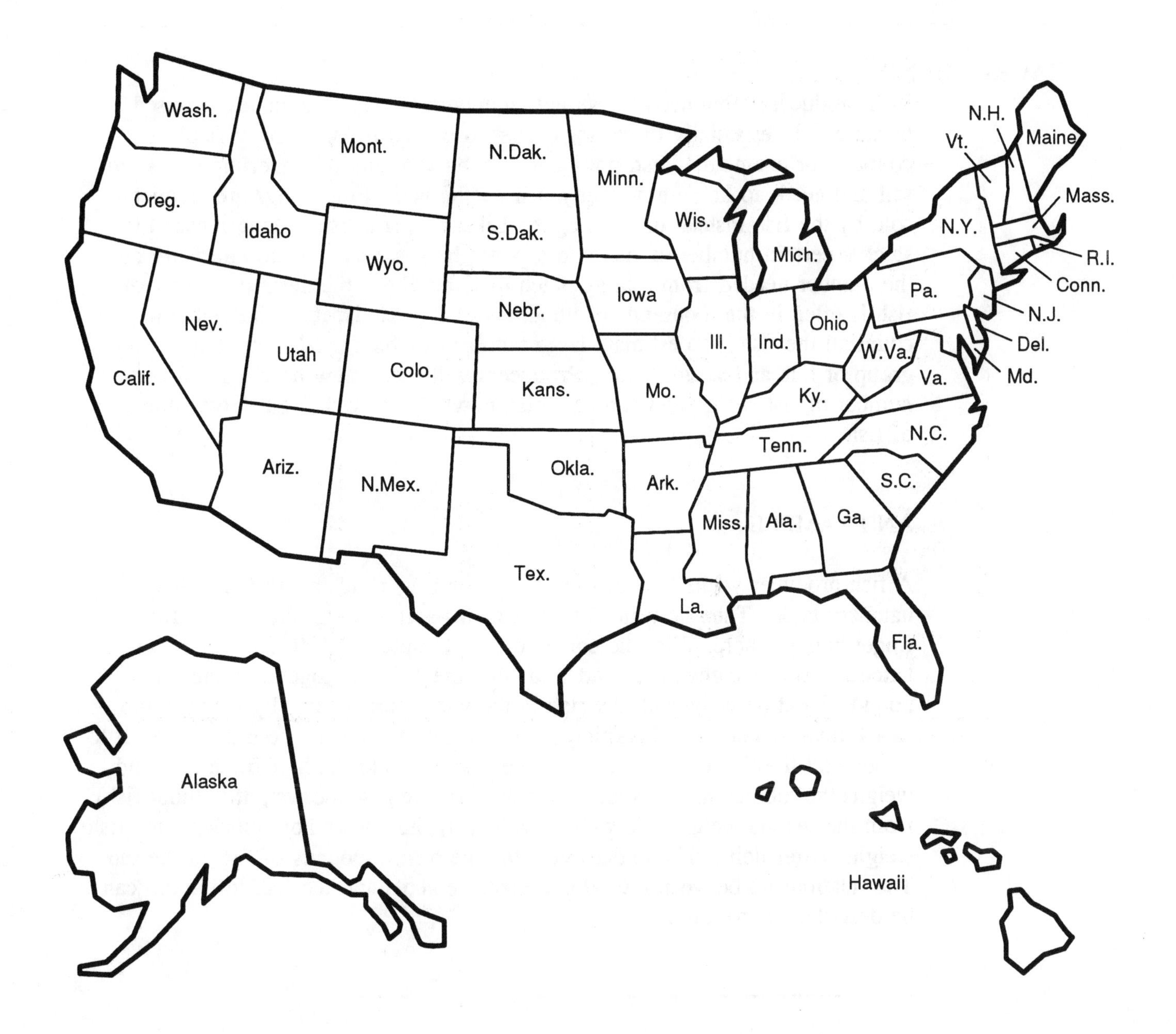
Wash.
Mont.
N.Dak.
Minn.
Oreg.
Idaho
S.Dak.
Wis.
Wyo.
Mich.
Nebr.
Iowa
Nev.
Utah
Colo.
Ill.
Ind.
Ohio
Calif.
Kans.
Mo.
Ky.
Tenn.
Okla.
Ariz.
N.Mex.
Ark.
Miss.
Ala.
Ga.
S.C.
N.C.
Tex.
La.
Fla.
Pa.
W.Va.
Va.
N.Y.
Vt.
N.H.
Maine
Mass.
R.I.
Conn.
N.J.
Del.
Md.
Alaska
Hawaii

Figure 4.3.2

NAME:______________________ DATE:____________ CLASS:____________

Chapter 5: The Aquaculture Industry

Laboratory Exercise 5.1 Estimating Fish Population

BACKGROUND

Fish producers often need to estimate numbers of fish. Estimates are used because fish, especially fingerlings (one- to two-inch fish), are difficult to count. For example, a fish producer may have a tank of fingerling catfish to sell and needs to know how many fish he/she has. Fingerlings are frequently sold by the fish instead of by weight. Likewise, if a fish producer wants to stock a certain number of fish into a pond, he/she may need to "measure out" the number needed from a large batch of fish. Since the desired number of fish is often in the thousands or hundreds of thousands, it is impossible to count all the fish. To estimate large numbers of fish, producers count a small group of fish and weigh them. Producers will then know how much a certain number of fish weigh and can use this knowledge to weigh out larger numbers of fish.

AN EXAMPLE

A fish producer wants to estimate how many fingerling catfish are in his hatchery tank. Then he wants to stock a two-acre pond at the rate of 1000 fingerlings per acre. First he counts out a "sample" of 100 fish from the hatchery tank, weighs them, and finds that the 100 fish together weigh 0.5 pound. Next he weighs all the fish in the vat. He does this by filling a tub half full with water and weighing the tub and water. Then he uses a dip net to remove all the fish from the tank, places the fish into the half-filled tub, and weighs the tub again. By subtracting the first weight (the weight without fish) from the second weight (the weight with fish), he knows how much all the fish weigh. After doing this he discovers that he has 30 pounds of fish in the tub. The relationship between the sample measurements and the fish in the tub can be described as follows:

$$\frac{\textit{Sample Number}}{\textit{Sample Weight}} = \frac{\textit{Total Number}}{\textit{Total Weight}}$$

Using this relationship, the fish producer cross multiplies and estimates that he has 6000 fish in the tub.

$$\frac{100\ fish}{0.5\ pounds} = \frac{?\ fish}{30\ pounds} \rightarrow \frac{100 \times 30}{0.5} = 6000\ fish$$

Next the producer wants to measure out enough fish to stock a two-acre pond at the rate of 1000 fish per acre. He multiplies his acreage by the desired number of fish per acre and figures out that he needs to measure out 2000 fish.

The producer does not want to count 2000 individual fish, so he will measure out approximately 2000 fish by weighing them. Using the following formula, he can find out how much 2000 fish will weigh.

He uses the same formula as before but substitutes 2000 into the total numbers portion of the equation and cross multiplies to calculate the weight of 2000 fish.

$$\frac{100\ fish}{0.5\ pounds} = \frac{2000\ fish}{?\ pounds} \rightarrow \frac{2000 \times 0.5}{100} = 10\ pounds$$

OBJECTIVES

- To estimate the number of fish in a group
- To estimate a certain number of fish from a large group

EQUIPMENT

one 8" X 8" baking pan
one 8" X 13" baking pan
laboratory scale
one quart jar

MATERIALS

one quart jar of navy beans

SAFETY

There are no special safety precautions associated with this exercise.

PROCEDURES

PART I

1. Beans will be used to illustrate how to estimate large numbers of fish - or other items that are not easily counted. In this exercise you will estimate the number of beans in a quart jar in the same manner that the fish producer estimated the number of fish in the tub.

2. Weigh an empty quart jar on the scale and record the weight in Table 5.1.1. Then fill the jar with beans and weigh the jar and beans. Record your answer in Table 5.1.1. Subtract the weight of the empty jar from the weight of the full jar to get the weight of the beans. Record your answer in Table 5.1.1 of the Results and Discussion section.

3. Count out 50 beans from the jar and weigh them. Now you know

   - the weight of all the beans
   - the weight of the sample of beans
   - the number of beans in the sample.

   With this information you can plug in the missing figures in the formula and solve it mathematically to estimate the number of beans in the jar. Complete item 2 in the Results and Discussion section.

PART II

4. In this part of the exercise the beans will represent fingerlings used to stock ponds. The baking pans will represent ponds. Imagine that an 8-inch by 8-inch baking pan represents a one-acre pond. Since an 8-inch by 8-inch pan contains 64 square inches, you can assume that every 64 square inches equals an acre. An 8-inch by 13-inch pan contains 117 square inches. Since 117 divided by 64 equals 1.8, you can say this pan represents a 1.8-acre pond. In this exercise you will "stock" the two pans at the rate of 300 beans (fish) per acre.

5. First you will stock the 8-inch by 8-inch pan. Since this pan represents exactly one acre you will need to weigh out 300 beans. Complete item 3 in the Results and Discussion Section.

6. Next you will stock the 8-inch by 13-inch pan. Since this pan represents 1.8 acres, you will need to weigh out 540 beans (1.8 acres times 300 beans per acre). Complete item 4 in the Results and Discussion section.

7. Now compare the two pans. Do the beans look like they are "stocked" at the same density in both pans? They should. If they do not look like the same densities, you may have made a math error. Double-check the math and try again. Complete Table 5.1.2 in the Results and Discussion section.

8. Count the beans in the two pans to see how close you came to the number you wanted. Record the results in Table 5.1.2.

RESULTS AND DISCUSSION

1. Complete Table 5.1.1 to determine the weight of one quart of beans.

Table 5.1.1 Determining the Weight of One Quart of Beans	
Weight of the jar with beans	
Weight of the empty jar	
Weight of the beans	

2. Complete this formula to calculate the number of beans in the jar:

$$\frac{\textit{Sample Number}}{\textit{Sample Weight}} = \frac{\textit{Total Number}}{\textit{Total Weight}} \rightarrow \_\_\_\_ = \underline{\ \ ?\ \ } \rightarrow \_\_\_\_\times\_\_\_\_ \rightarrow \textit{beans}$$

There are ________ beans in the jar.

3. Use the formula given here to help calculate how much 300 beans weigh. Record the result in Table 5.1.2.

$$\frac{\textit{Sample Number}}{\textit{Sample Weight}} = \frac{\textit{Total Number}}{\textit{Total Weight}} \rightarrow \frac{\_\_\_\_}{\_\_\_\_} = \frac{\_\_\_\_}{?} \rightarrow \_\_\_\_\_\_ \times \_\_\_\_\_\_ \rightarrow \textit{(weight)}$$

Slowly add beans to the scale until it reads the weight calculated in the formula. Now spread these beans evenly in the bottom of the 8-inch by 8-inch pan.

4. Use the formula given here to calculate how much 540 beans weigh. Record the result in Table 5.1.2.

$$\frac{\textit{Sample Number}}{\textit{Sample Weight}} = \frac{\textit{Total Number}}{\textit{Total Weight}} \rightarrow \frac{\_\_\_\_}{\_\_\_\_} = \frac{\_\_\_\_}{?} \rightarrow \_\_\_\_\_\_ \times \_\_\_\_\_\_ \rightarrow \textit{(weight)}$$

Slowly add beans to the scale until it reads the weight calculated in the formula. Now spread these beans evenly in the bottom of the 8-inch by 13-inch pan.

5. Record the results of Part II (the pond stocking exercise), in Table 5.1.2.

Table 5.1.2 Stocking Rate of Fingerlings in Ponds				
Pan Size	Represents Pond Size	Recommended Stocking Rate for this Pond	Weight Needed	Actual Count
8" X 8" pan	1 acre	300		
8" X 13" pan	1.8 acres	540		

6. Do the pans look like they have the same density of beans?

7. Based on your results in this exercise, do you think estimating numbers by weight is a reliable method? Explain.

8. Name other situations in which this method of estimating would be useful.

NAME:______________________________ DATE:______________ CLASS:______________

Chapter 5: The Aquaculture Industry

Laboratory Exercise 5.2 The Oxygen Content of Water

BACKGROUND

Fish extract oxygen dissolved in water through their gills. If there is not enough oxygen, fish can become stressed or even die. Low oxygen levels are a common cause of fish kills in ponds. The amount of dissolved oxygen has other indirect, but very important, influences on fish health because:

1.) Chemicals such as carbon dioxide, hydrogen sulfide, and nitrites are common fish toxins found in ponds. Fish are better able to tolerate these toxins if they have plenty of oxygen available to them. For example, carbon dioxide is not very toxic to fish if the oxygen level is high but can be deadly if the oxygen level is low.

2.) Oxygen can also reduce the concentration of some toxins by chemically reacting with them and converting them to nontoxic chemicals.

The fish species and size being raised, water temperature, concentration of other chemicals present, and other factors determine what level of oxygen is necessary. To demonstrate how complex oxygen needs are, consider the following points:

- Species such as trout require more oxygen than catfish or carp.
- Large fish need more oxygen than small fish of the same species.
- Since fish are cold blooded, their metabolism and need for oxygen goes up as the water temperature rises, yet warm water holds less oxygen than colder water.
- Fish need more oxygen when carbon dioxide levels are high than when they are low.

Even well-educated fish producers cannot take all these factors into account. Instead producers use rules of thumb determined from experience and research. Generally, prolonged exposure to oxygen levels below 1 part per million (PPM) will kill fish, especially larger fish. If fish are exposed for prolonged periods to oxygen levels between 1 and 5 PPM, they will become stressed and growth will be slow. Oxygen levels above 5 PPM are desirable.

In this exercise you will use a commercial test kit to measure the dissolved oxygen in samples of pond water and tap water. This is the same procedure used by industry professionals.

OBJECTIVES

- To test pond water for dissolved oxygen levels
- To determine if the oxygen level would be acceptable for fish to live and grow
- To determine the difficulty of adding oxygen to water by mechanical agitation

---

EQUIPMENT

commercial water test kit

sources of water test kits:

Hach Company	LaMotte Company
P.O. Box 389	P.O. Box 329
Loveland, CO 80539	Chestertown, MD 21620

quart jar

MATERIALS

samples of pond or lake water

SAFETY

The water test kit has a section on safety included in the direction book. Be sure to read the safety section before beginning this exercise.

PROCEDURES

1. Obtain the oxygen test kit and a sample of pond/lake water from your teacher. Follow the instructions included in the kit for testing dissolved oxygen. Complete item 1 in the Results and Discussion section.

2. Test the dissolved oxygen in a sample of tap water. Record the results in Table 5.2.1 in the Results and Discussion section.

3. Air pumps are used in aquariums, and large aerators that churn the water are used in ponds, to help dissolve oxygen into the water. Use the following procedure to determine if you can increase the dissolved

oxygen content in the tap-water sample. Place a sample of tap water in a container and tighten the lid. Shake vigorously for at least five minutes.

4. Rerun the dissolved oxygen test on the tap water after shaking. Complete Table 5.2.1 in the Results and Discussion section.

RESULTS AND DISCUSSION

1. What is the dissolved oxygen level of the pond water sample?________

   Would this level be adequate for the production of channel catfish? ____

   Would the oxygen level be good for raising trout? ________

2. Complete Table 5.2.1.

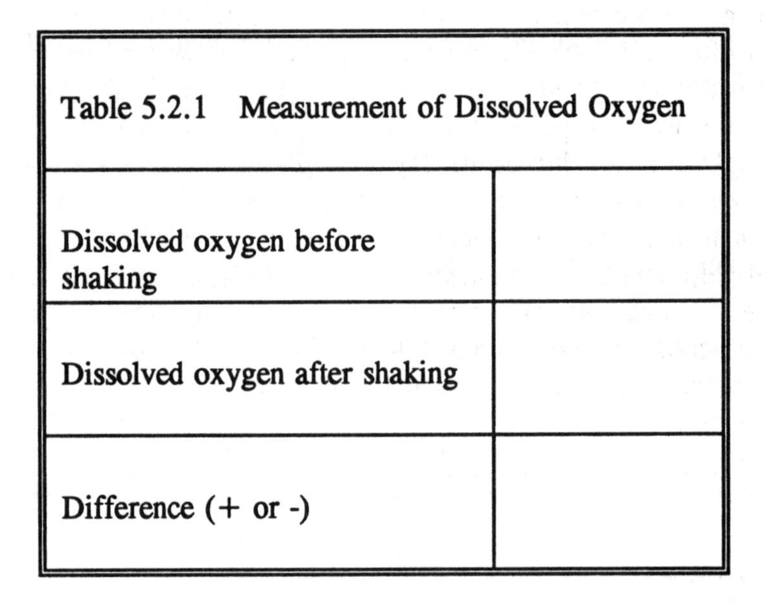

Table 5.2.1 Measurement of Dissolved Oxygen	
Dissolved oxygen before shaking	
Dissolved oxygen after shaking	
Difference (+ or -)	

3. Were you able to increase the oxygen level in the tap water sample by shaking? If not, how long do you think you would have to shake to detect a difference in the oxygen level?

4. Based on your experience in adding oxygen to water by shaking, how effective do you think mechanical agitators are in ponds?

NAME:______________________ DATE:______________ CLASS:____________

Chapter 5: The Aquaculture Industry

Laboratory Exercise 5.3 Testing Water pH in Ponds and Lakes

BACKGROUND

Like all animals, fish need a healthy environment to grow and prosper. Maintaining good water quality is essential to assure good growth and to show a profit. Chemical compounds such as hydrogen sulfide gas, ammonia, nitrites, and carbon dioxide can stress or kill fish if concentrations become too high in the culture pond. Water chemistry can become very complex and confusing to the average fish producer. But in most cases if the producer can maintain an adequate water pH and dissolved oxygen levels, most of the other water chemistry parameters need not be of particular concern.

The pH of water is also important to the fish producer. The desirable range of pH in fish ponds is between 6.5 and 9. A pH lower than 4 or higher than 11 will often kill fish. During photosynthesis, aquatic plants remove CO_2 from the water and the pH rises. At night, respiration in the pond produces CO_2, which causes the pH to decrease. As you might expect, pH tends to go down at night and rise during the daylight. Basic ions in water can reduce the degree to which pH fluctuates during a day. Alkalinity is a measure of the amount of bases in the water. If adequate bases are present in the water, they will help counteract the acidic effects of CO_2, and the pH will not fluctuate as much. Alkalinity can be increased with the addition of agricultural limestone. As a general rule of thumb, a total alkalinity level of twenty parts per million or higher is desirable for good fish growth.

OBJECTIVES

- To test water pH
- To determine if the pH level is suitable for good fish growth
- To test total alkalinity in pond water and determine if lime is needed

---

EQUIPMENT

commercial water test kit

MATERIALS

samples of pond, lake, or stream water
tap water
distilled water

SAFETY

The water test kit has a section on safety included in the direction book. Be sure to read the safety section before beginning this exercise.

PROCEDURES

1. Obtain a commercial water test kit and one or more samples of pond, lake, or stream water from your teacher. Follow the instructions included in the kit for testing pH. Complete Table 5.3.1 in the Results and Discussion section.

2. Test the pH of a sample of distilled water. Record the results in Table 5.3.1 in the Results and Discussion section.

3. You may also want to test the pH tap water, but the presence of chlorine in the water may interfere with the test results. Some test kits explain how to remedy this problem. Most of the chlorine can be eliminated by letting the water sit out in an open container for twenty-four hours. Also, pet shops sell a chemical to put in aquarium water to eliminate chlorine.

RESULTS AND DISCUSSION

1. Why does the pH of pond water fluctuate day to night?

2. What pH level is desirable for ponds and lakes? What may happen if the pH level is below 4 or above 9?

3. Would the pH of a pond be higher on a sunny day or on a cloudy day? Why?

4. Complete Table 5.3.1.

Table 5.3.1 pH of Water Samples

Sample Number	Source	pH	Lime Needed?
1			
2			
3			
4	Distilled water		

5. If a pond is muddy and there are no plants in it, would the pH change throughout the day? Why?

6. Which pond would probably have a higher pH, one with a limed pasture for a watershed or one with a forested watershed?

NAME:________________________ DATE:____________ CLASS:____________

Chapter 6: Alternative Animal Agriculture

Laboratory Exercise 6.1 Producing Mealworms

BACKGROUND

Fish bait is probably the most commonly mentioned alternative animal in agriculture. Many people make a living from raising or selling worms to fishers, pet shops, pet owners, etc. Earthworm castings are especially valuable in improving the soil. In fact the ability of earthworms to improve the soil is much more valuable to agriculture than sales of the worms themselves.

The two most commonly used worms for fish bait are the earthworm and the mealworm. Either one may be raised easily in the classroom. Mealworms, also called golden grubs, have been selected for this exercise because they are cleaner and are odorless. The disadvantage is that their life cycle is relatively long.

Many insect larvae are called mealworms, even the maggots of certain flies. The mealworm you will be working with is a beetle in adulthood. The larval stage is best known. It is a golden-yellow larva of the darkling beetle. It is a common pest of stored grain. Unfortunately they can also escape into kitchens and infest your oatmeal or other food.

Mealworm Life Cycle

Mealworms are not true worms but are the larval form of the darkling beetle. This beetle undergoes complete metamorphosis which means their life cycle consists of egg, larva, pupa, and adult stages. The eggs are very small and white in color. The eggs incubate for one to two weeks before hatching into small larvae. The larvae are called mealworms and grow for several months before transforming into the pupa stage. During the one to two weeks of the pupa stage, the mealworm transforms into the adult beetle. The adult beetle lives for only about a month. During this time, a single female may lay as many as five hundred eggs before dying. The entire life cycle of the beetle lasts about six months. Temperature greatly affects the development time of the beetle. Cooler temperatures tend to slow down development time, while warmer temperatures tend to speed up development time.

The mealworm colony should be kept in a warm environment. Cool temperatures delay their development, while warmer temperatures speed the process.

OBJECTIVES

- To distinguish each stage in the metamorphosis of a mealworm
- To produce an alternative agricultural animal

EQUIPMENT

*small container such as a pint or quart margarine tub or coffee can with lid
one nail
knife
shallow tray
one black sheet of paper

* Glass jars can also be used, but worms raise best in containers with a lot of surface area rather than deep containers.

MATERIALS

one cup of chick starter mash, wheat bran, corn flakes, or other dry breakfast cereal
one quarter of a medium-sized white potato
two tablespoons of four different feeds

SAFETY

Safety glasses should be worn by the user and observers if a hammer and nail are used to punch holes in lids. Mealworms and the adults are completely harmless. They are completely safe to handle.

PROCEDURES

Part I

1. Punch ten to fifteen air holes in the container lid with a nail.

2. Put one cup of chick starter mash, corn flakes, or other feed material in the jar.

3. Cut off 1/4 of a medium-sized white potato and place it on top of the feed with the cut side down. Mealworms can live, grow, and reproduce without any added moisture, but they reproduce faster if

moisture is supplied occasionally in the form of slices of potato, slice of carrot, etc.

4. Obtain a sample population of mealworms from the teacher. You should obtain a mixture of adults and larvae. Your teacher may assign a certain number of each. Place the mealworms in the container. With a mixed population you should see new larvae emerging in a little over two weeks.

5. Cover the entire surface with a paper towel to provide a secluded environment.

6. Place the lid on the container. The lid is not usually necessary, although occasionally an adult will crawl out. The lid does prevent spillage and allows several colonies to be stacked in the classroom. Many mealworm colonies are kept in open top plastic pans.

7. Place the colony in an area of the classroom out of direct sunlight. Leave it undisturbed for at least two weeks.

8. More feed and potato slices should not be needed for several weeks, but the colony should be monitored in case the food supply is depleted.

9. At a time designated by the teacher, the colony may be examined and a count of the worms made.

10. About every three to six months the colony should be rejuvenated by removing the adults and worms, discarding the other contents of the jar, and adding fresh feed.

Part II

11. Place a mealworm larva and adult in a petri dish. Examine them for body segments and legs. Complete item 1 in the Review and Discussion Section.

12. Determine which feed mealworms prefer. In a large flat container place four different types of food. Place several larvae in the center. Carefully examine each container the next day to determine in which feed the most larvae are found. Complete the chart in number 3 in the Results and Discussion section below.

13. Cover the bottom of a tray with a light covering of feed. Cover one-half of the tray with a black piece of paper. Place ten mealworms in the center of the tray. Observe to determine where the most worms move to in the tray.

RESULTS AND DISCUSSION

1. Complete Table 6.1.1 over the duration of the mealworm study as directed by your teacher.

Table 6.1.1 Mealworm Colony Record

	Number at Establishment	After 3 Weeks	Percent Increase	After 6 Weeks	Percent Increase from Establishment	After 9 Weeks	Percent Increase from Establishment
Number of adults							
Number of larvae							
Number of pupae							

2. Record in Table 6.1.2 the number of body segments and appendages observed on the larvae.

Table 6.1.2 Mealworm Observation Form

	Number of Body Segments	Number of Legs
Adult		
Larvae		

3. How many worms moved to the light area of the tray? What is your conclusion?

4. How many worms moved to the dark end of the tray? What is your conclusion?

NAME:____________________ DATE:______________ CLASS:______________

Chapter 7: The Classification of Agricultural Animals

Laboratory Exercise 7.1 Classifying Animals

BACKGROUND

You use classification systems in everyday life. You are constantly being classified as well. For example, you are in a school grade based on common characteristics - age and ability. Other examples are clubs, where you probably classed yourself by common interest, and the telephone book listing, where you are categorized by last name and first name.

Animals and other living things in the world have been identified, grouped, and classified in an attempt to more effectively study and communicate about them. Plants, animals, and other organisms are classified or grouped together by characteristics they have in common.

Animals are given names according to a scientific classification system. This system was developed by a Swedish botanist named Carolus Linnaeus who grouped organisms according to similar characteristics. Linnaeus used two Latin names for identifying each individual organism. The first of the two names is the **genus** and the second part of the name is the **species.**

In this laboratory exercise you will identify the common characteristics and differences in selected animals. This is the first step in the classification process.

Linnaeus's Contribution to Science

Linnaeus gave scientists an orderly system with which to classify living organisms. Scholars recognized the value of such a system, and it was soon used world-wide. The descriptive method, standard names, and practical method devised by Linnaeus laid the foundation for modern scientific research. The system is still in use today.

OBJECTIVES

- To identify characteristics that can be used to group objects
- To identify common characteristics of several agricultural animals
- To explain the basic concepts of a classification system for animals

EQUIPMENT

No special equipment is necessary for this exercise.

MATERIALS

Text - *The Science of Animal Agriculture*

SAFETY

No special safety precautions are necessary for this exercise.

PROCEDURES

1. Observe the cow, bee, chicken, and pig in Table 7.1.1 in the Results and Discussion section. Complete the table by listing all the similarities of these animals.

2. In Table 7.1.2, list the characteristics that put the animals pictured into distinct subgroups from each other. List all characteristics of that animal that are not characteristics of the other three. Use the text, *The Science of Animal Agriculture*, as necessary.

RESULTS AND DISCUSSION

1. Why are plants and animals classified?

2. List types of classifications that you have been put into (class, etc.).

3. What are the two parts of scientific names and what language are they written in?

4. In the space provided, list all the similarities and differences in the animals pictured in Tables 7.1.1 and 7.1.2.

Table 7.1.1 Classification by Similarities

Cow	Bee
Pig	Chicken

List similarities:

Table 7.1.2 Distinguishing Characteristics

Cow	Bee
Distinguishing characteristics:	Distinguishing characteristics:
Pig	**Chicken**
Distinguishing characteristics:	Distinguishing characteristics:

NAME:________________________ DATE:______________ CLASS:______________

Chapter 7: The Classification of Agricultural Animals

Laboratory Exercise 7.2 Classification systems

BACKGROUND

Scientists often use a "key" to help classify and identify animals. A key is a listing of characteristics, such as number of legs, type of skeleton, and type of skin covering. Keys are especially useful in agricultural applications in classifying insects and diseases.

The first step, and one of the most important steps in controlling a pest, is to identify it. Environmentally conscious (and cost conscious) producers do not guess at the type of pest and apply pesticides indiscriminately. If identification cannot be made, the pest is usually sent to a higher authority for assistance. This person may be a scientist at a university or at a private laboratory. The scientist may be able to identify the pest on sight; but if not, he/she may use a key to assist in its identification.

Keys start by listing broad characteristics of plants and animals. From there they branch to continuously more-specific characteristics. You begin at the first statement and follow directions to trace the characteristics of the plant or animal to a specific category or name.

In this laboratory you will "key-out" common objects to simulate the classification of animals or plants. Rather than bring several kinds of animals to the classroom, you will learn the concept of the use of a key by using common objects. You will use pencils, pens and, markers in this simulation. To classify objects you must be able to examine them in detail.

OBJECTIVES

- To identify characteristics that can be used to group objects
- To explain the basic concepts of a classification system for animals

---

EQUIPMENT

A variety of writing utensils - one of each type pictured in Figure 7.2.1 will be ideal. The teacher may direct you to work from Figure 7.2.1.

MATERIALS

No special materials are needed for this laboratory exercise.

SAFETY

No special safety precautions are necessary for this laboratory exercise.

PROCEDURES

1. Obtain a collection of writing utensils similar to those shown in Figure 7.2.1. If these are not available, the photograph of these items may be used. Lay the items out as shown.

2. See if you can follow the key in Table 7.2.1 to find the name of each item. For this simulation the names of the items are the names of the companies that produced that product.

3. Select one item. Read statement #1 of the key and select the statement that is true for that item. Follow the directions to trace this item to the proper name.

4. Repeat step 3 to identify each item. Write the name of each item beside its picture in Figure 7.2.1.

Table 7.2.1		Key to Identifying Writing Tools	
1.	A. Uses ink		Go to 2
	B. Uses lead or solid core		Go to 3
2.	A. Has a pocket clip		Go to 5
	B. Does not have a pocket clip		Go to 6
3.	A. Has an eraser		Go to 4
	B. Does not have an eraser		Go to 8
4.	A. Is sharpened in a pencil sharpener		Go to 10
	B. Lead is exposed by peeling away covering		Go to 8
5.	A. Has a non-removable pocket clip		Name: Fisher
	B. Has a pocket clip on a removable cap		Go to 7
6.	A. Ink flows freely from a felt tip		Go to 9
	B. Ink is dispersed from a ball point		Name: Paper Mate
7.	A. Makes very wide marks		Name: Phoenix
	B. Makes narrow marks		Go to 6
8.	A. Lead has a grease base		Name: Chinon
	B. Lead is not grease based		Name: Unix
9.	A. Makes very wide marks		Name: Dennison
	B. Makes narrow marks		Name: Bic
10.	A. Is a #2		Name: Officemate
	B. Is not a #2		Name: Genie

RESULTS AND DISCUSSION

1. Use the key in Table 7.2.1 to determine the names of the writing utensils shown in Figure 7.2.1. Write the names below each item.

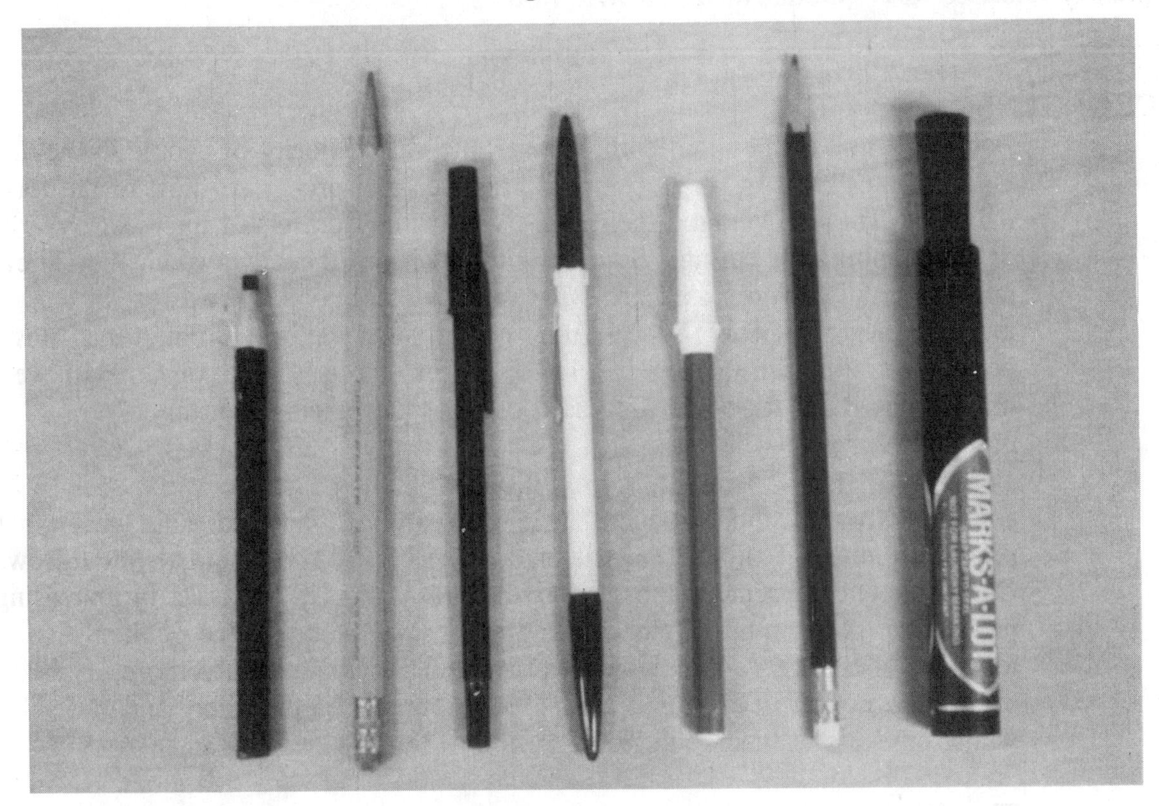

Figure 7.2.1 Write the name of each item beside its picture

2. List the distinguishing characteristics that were used in classifying the items in Figure 7.2.1.

3. Why are keys to identification important tools?

4. Explain how an identification key works.

NAME:________________________ DATE:____________ CLASS:____________

Chapter 8: Consumer Concerns

Laboratory Exercise 8.1 Consumer Confidence

BACKGROUND

Before the end of the nineteenth century, the vast majority of people in the United States lived in rural areas and produced and processed most of the food they ate. They slaughtered their own livestock and preserved the meat by drying, canning, or curing. Since people processed their own food, they knew how the food was processed and what went into the food. However, in modern times few people process their own food. The only product that most people see is the finished product in the grocery store. Most consumers have no idea how their food was processed or what ingredients went into the product.

Consumers want a food product that is relatively inexpensive. This means that producers must get as much efficiency as possible from the animals they grow. Chemicals and substances approved by the FDA and USDA assist in increasing production. Most modern techniques are not understood by the public. As technology increases and new discoveries help growers produce more efficiently, concerns are raised among consumers as to the safety and wholesomeness of their food.

The USDA inspects all meat products processed and sold in the United States for human consumption. The purpose of meat inspection laws is to ensure that the very best, most wholesome meat reaches the consumer. Meat inspection guarantees that the meat will be safe, wholesome, and accurately labeled.

Still, consumers worry. Sporadic outbreaks of illnesses due to contaminated meat products make national news. It is difficult for the public to understand the highly technical system of livestock production and meat processing, and people tend to be suspicious of things they do not understand.

In this laboratory exercise you will collect data using a questionnaire. You will draw conclusions as to how people feel about some of the most controversial topics in animal production and food processing.

OBJECTIVES

- To determine consumer concerns with various phases of the animal and meat industry

- To demonstrate the use of survey techniques to determine public opinion
- To demonstrate the use of research data to draw conclusions

---

EQUIPMENT

No special equipment is necessary for this laboratory exercise.

MATERIALS

Table 8.1.2, Consumer Concerns Questionnaire
Text - *The Science of Animal Agriculture*

SAFETY

No special safety precautions are necessary for this laboratory exercise.

PROCEDURES

1. Review chapter 8 from the text *The Science of Animal Agriculture* and the questionnaire in Table 8.1.2. Make sure you understand each question.

2. Ask members of the community to participate in a survey using the questions in Table 8.1.2. Tell them that it will take no more than five minutes.

3. Read each question aloud to the participants and ask them to rank the item 1 to 4. See Table 8.1.1 Record their answers as directed in item 2 of the Results and Discussion section.

Table 8.1.1 Questionnaire Rating Scale
"No Confidence" = 1
"Fairly Confident" = 2
"Confident" = 3
"Highly Confident" = 4

4. Complete the items in the Results and Discussion section.

RESULTS AND DISCUSSION

1. Briefly outline what *The Science of Animal Agriculture* says about these consumer concerns.

The safe handling and inspection of meat products...

The safety of medications administered to animals slaughtered for meat...

The safety of hormones administered to animals slaughtered for meat...

The danger from cholesterol in pork and beef...

The dangers genetic engineering may pose...

How animal production may affect the environment...

2. Record the responses from those surveyed in Table 8.1.2. Place a tally mark in the cell that corresponds to the response. Since you will be reading the items aloud and recording the responses, all responses may be put on one form.

3. Average the responses across the rows and record this figure in the column labeled "average." Determine the average by assigning a number to each response. Assign numbers as follows:

   "No Confidence" = 1

   "Fairly Confident" = 2

   "Confident" = 3

   "Highly Confident" = 4

4. Use the average of the responses to draw conclusions about how consumers feel about each category. Write a short paragraph about your conclusion in the space provided.

   The safe handling and inspection of meat products...

   The safety of medications administered to animals slaughtered for meat...

   The safety of hormones administered to animals slaughtered for meat...

   The danger from cholesterol in pork and beef...

   The dangers genetic engineering may pose...

   How animal production may effect the environment...

TABLE 8.1.2 Consumer Concerns Questionnaire

Topics and Questions	(1) No Confidence	(2) Fairly Confident	(3) Confident	(4) Highly Confident	Average for Each Row
Meat Products					
1. How confident are you that the meat industry is supplying wholesome meat for the public?					
2. Are you confident that processors handle meat safely and sell a healthy product?					
3. How confident are you that USDA meat inspection is adequate?					
4. How confident are you that your supermarket does a good job of handling fresh meat and sells a healthy product to you?					
5. How confident are you that labels on meat products are accurate?					
Medications					
6. How confident are you that medications fed to livestock are safe?					
7. How confident are you that medications given to animals do not cause disease organisms to become resistant to the medication, rendering the medicine ineffective in humans?					
8. How confident are you that medications fed to animals are not passed on to humans?					
Hormones					
9. How confident are you that hormones used in animals do not pose a threat to humans?					

Chapter 8: Consumer Concerns, Exercise 8.1

TABLE 8.1.2 Consumer Concerns Questionnaire (continued)

Topics and Questions	(1) No Confidence	(2) Fairly Confident	(3) Confident	(4) Highly Confident	Average for Each Row
10. How confident are you that hormones used in animals are not passed on to humans through meat and milk?					
Cholesterol					
11. How confident are you that pork and beef are healthy foods?					
12. How confident are you that cholesterol from pork is not a major contributor to heart disease?					
13. How confident are you that cholesterol from beef is not a major contributor to heart disease?					
Genetic Engineering					
14. How confident are you in the general safety of genetic engineering?					
15. How confident are you that genetic engineering will greatly improve the health of animals and humans?					
Environmental Concerns					
16. How confident are you in the EPA to regulate the safe disposal of animal wastes?					
17. How confident are you that your water supply is safe from farm animal pollutants?					
18. How confident are you that farm animals are not a major contributor to global warming?					

NAME:____________________ DATE:__________ CLASS:__________

Chapter 9: The Issue of Animal Welfare

Laboratory Exercise 9.1 Animal Rights - Pro and Con

BACKGROUND

"Rabbits have been used in a torturing experiment in which bristles have been shoved into their eyes."

"Some researchers torture animals by placing them on a hot plate just to observe their reaction."

"In one experiment, researchers pinch tender points on the animal's body with such devices as alligator clips."

"Farmers don't care about their animals. Farm animals are often beaten and generally mistreated."

"Grain fed to farm animals for slaughter could be shipped to starving countries if Americans would only stop eating meat."

The statements above are typical of points made by members of the animal rights movement. These statements may evoke an emotional response from you, either for or against their argument. Such statements made to the public can be very detrimental to scientific research and agricultural production. Without an opportunity to hear the other side of their argument, animal rights activists can be very convincing.

Figure 9.1.1 Should researchers experiment on animals so that humans may live?

In this laboratory exercise you will read both sides of the argument. One side is that of the animal rights activists and the other is that of the agricultural producers and research scientists. In the Results and Discussion section you will be asked to state your personal position on these arguments and your reasons for choosing this position.

OBJECTIVES

- To state an opinion concerning the arguments of animal rights activists
- To analyze arguments for and against animal research and agricultural production of animals

EQUIPMENT

No special equipment is needed for this laboratory exercise.

MATERIALS

Written discussions between animal rights activists and rebuttals to their statements (included at the end of the Results and Discussion section).

SAFETY

There are no special safety precautions necessary with this exercise.

PROCEDURES

1. Read each statement in "Animal Rights Arguments and Rebuttal" as presented here.
2. After reading the entire article, reread each argument and its rebuttal. Then describe your feelings about the topics by answering the questions in the spaces provided in the Results and Discussion section.

RESULTS AND DISCUSSION

1. Should animals be used in medical research? State your opinion and explain why or why not.

2. In what type of research, if any, would you oppose using animals? Why?

3. Do animals have the same rights as humans? Explain your answer.

4. Do you think animals should be used in pain research or should humans only be used? Explain.

5. Why do some scientists consider pain research data collected from animals to be more accurate than data from humans?

6. What was your initial reaction to reading the introductory quotes from animal rights activists in the Background section of this exercise?

7. Why is it important for researchers and agricultural producers to counter the arguments of animal rights activists?

8. Why would it be to the advantage of agricultural producers not to mistreat their animals?

9. People who disagree with the animal rights activists say that even if animal production were halted, there would continue to be just as much hunger in the world. Do you agree or disagree? Why?

ANIMAL RIGHTS ARGUMENTS AND REBUTTAL

Animal Rights Activist:
The use of animals in research entails a great deal of suffering for the animals. Research procedures used on animals include a grisly array of painful procedures. In one experiment, researchers pinch tender points on the animal's body with such devices as alligator clips.

Rebuttal:
Most pain research in animals is of brief duration and ceases when the animal gives any indication of feeling pain. It is of no value to test their ability to tolerate continuous and high levels of pain. Alligator clips are used in surgery for clamping off delicate blood vessels and exert only very mild pressure. They can be clamped to most parts of the human body without causing significant pain. Animals may try to remove the clips, but it is normal for an animal to try to remove any object attached to them, painful or not.

Animal Rights Activist:
Animals are often subjected to barbaric torture. In one research study, bristles were shoved into the eyes of rabbits. Can you imagine the pain the animals had to endure? What could possibly be gained form such a torturing experiment?

Rebuttal:
This procedure is to test local anesthetics. It is a completely humane test that does not cause pain. The same test is used in humans. Fibers are touched to the skin or eye with increasing fiber diameter until a response is obtained - a blink of the eye. Rabbits are especially useful for studies of ability to sense stimuli to the eye because they sit for many minutes without blinking.

Animal Rights Activist:
Some researchers torture animals by placing them on hot surfaces to observe their reaction. In one experiment, rats were placed on a hot plate. Such inhuman treatment should never be allowed, and if it is, it should be confined to human volunteers.

Rebuttal:
Hot plates have been used to analyze the ability of animals to perceive a stimulus as painful. The hot plate is a surface heated to about 131°F - cooler than many street surfaces in the summer. A person could place his/her hand on this surface for several seconds before realizing the surface is uncomfortable. Rats lick their feet when the surface becomes hot, at which time they are removed from the surface.

Animal Rights Activist:

Humans, not animals, should be used exclusively in pain research. At least humans can tell the researcher their feelings. Is it right for one animal to endure pain for another animal (humans)? They have the same rights as humans.

Rebuttal:

Pain research in humans is extremely valuable, but experiments with animals can often provide much more accurate measurement. Humans understand that they are involved in an experiment and often endure much of the pain before reacting. Animals do not endure pain, they signal immediately when they feel the stimulus. Useful results can only be obtained from healthy animals without stress.

Animal Rights Activist:

Drugs being developed to benefit humans should be tested on humans. Why should animals be made to bear the pain of drug side effects?

Rebuttal:

Initial testing of new drugs should not be carried out in humans. Until the effectiveness of a drug and its potentially toxic effects have been determined, most people agree that it is much better to test it on rats rather than on humans.

Animal Rights Activist:

Farmers do not care about the welfare of their animals. Farm animals are often beaten and generally mistreated. In most modern production systems animals are confined to tiny cages and are not allowed to move freely. For example, sows are kept in stalls barely wide enough for their bodies. They cannot even turn around. Barns are often overcrowded, unventilated, and generally a miserable environment for animals.

Rebuttal:

A producer would be foolish to mistreat his/her animals. Healthy and happy animals are more productive. The love of animals is a major reason for people entering agriculture. There are good reasons to treat animals humanely and provide a healthy environment, but there are no advantages to mistreating animals or keeping them in unhealthy environments. Agricultural producers are like most of the general public and feel a moral obligation to avoid cruelty to animals. Animal pens are designed for the welfare of the animal and efficiency of production. A sow ready to give birth is kept in a narrow stall to prevent injury to the pigs. In an open pen many pigs are stepped on or crushed to death by the sow.

Animal Rights Activist:

Grain fed to animals for slaughter could be shipped to starving countries if Americans would only stop eating meat. This would save the animals and humans alike. The grain fed to livestock could feed the entire population of the earth. Why should animals be fed when people go hungry?

Rebuttal:

The public must realize that farmers can only afford to produce grain if there is a market - someone willing to purchase it at a fair price. If all livestock production operations were shut down, who would purchase the grain to ship to poor countries? Who would pay? The poor countries cannot afford to pay. If they could, their people would not be starving. Would the American public be willing to pay? Probably not. If they were, we could be sending food to those countries today. The American farmer has, and has had, the capacity to produce a lot more grain than is currently being produced. If someone wanted to buy it, farmers could produce enough grain to feed the world and livestock.

NAME:______________________ DATE:______________ CLASS:____________

Chapter 10: Animal Behavior

Laboratory Exercise 10.1 As the Animal Sees It

BACKGROUND

The scientific study of agricultural animal behavior is relatively new. One reason for its late development may be that it is a difficult field of study. Think of the problems a scientist might face when trying to determine what an animal sees and hears, how it interprets what it sees and hears, and how it reacts to these stimuli. Since animals cannot tell us directly, we have to depend on observation of the actions of the animals and interpretation of these actions.

A better understanding of animal senses and their degree of development will make producing, handling, and marketing livestock easier. For example, here are some common happenings that most people never think about or associate with animal reactions.

1. A piece of paper blowing in front of cattle may spook them and make them turn from the direction you are driving them.

2. Sunlight, creating glare in some spots and dark shadows in others on the walls of a loading chute, may make animals balk or turn back.

3. Holes/trenches in the ground, such as a cattle guard, will make cattle turn back. It is said that black strips painted on the ground elicits the same response - probably because the stripes look like holes to a cow.

Because of the difficulty in setting up an animal experiment in the school setting, we will use human behavior to try to understand just how much we do not yet know about animal behavior. Many of us do not understand some of our own actions or perceptions. This is especially true with sight. Optical illusions can play tricks on our minds.

In this exercise you will perform three short experiments with the human eye. You may be surprised at the results. When you see how easily the human eye can be fooled, you may understand better why animals are sometimes spooked at the slightest movement.

OBJECTIVES

- To explain that animals may react to stimuli that seem trivial to humans
- To define an optical illusion
- To infer how unusual stimuli and surroundings may affect animals

---

EQUIPMENT

crayons or colored pencils
one ruler

MATERIALS

one white sheet of paper
two coins
one index card

SAFETY

No special safety precautions are necessary with this exercise.

PROCEDURES

Part I

1. Place two coins between your thumb and forefinger, as shown in Figure 10.1.1.

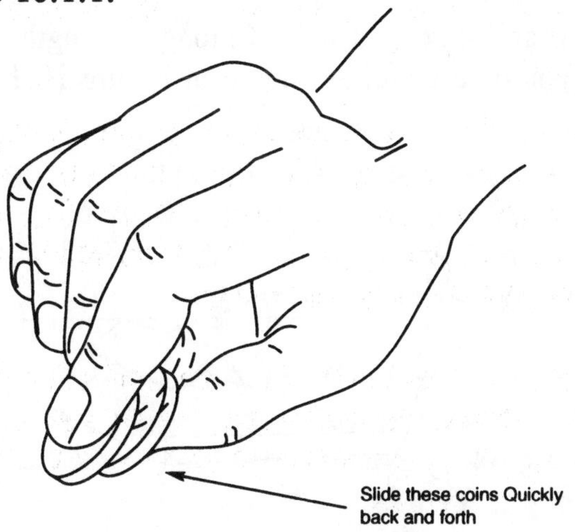

Figure 10.1.1 Rub the coins rapidly between your thumb and forefinger

2. Rub the coins back and forth between your fingers as fast as you can. Look at the coins as you do this. When you see three coins, you have perfected this exercise.

3. While rubbing the coins, show them to students who have not observed your preparation. Ask how many coins you have. If they say three, drop the two coins in their hands to show that there are only two.

4. Answer item 1 in the Results and Discussion section concerning the coin exercise.

Part II

5. Observe the hat in Figure 10.1.2. Is it taller than it is wide or wider than it is tall? Complete item 2 in the Results and Discussion section.

Figure 10.1.2 Is the hat taller than it is wide or wider than it is tall?

Part III

6. Fold an index card in half along its length and place it directly in front of you on a table, as shown in Figure 10.1.3.

Figure 10.1.3 Place the folded index card in front of you as shown

7. Determine your dominant eye by holding a pencil as shown in Figure 10.1.4. With both eyes open, line up the point of the pencil with the corner of the room. Close one eye and look at the pencil. Repeat the procedure with the other eye. The eye through which the pencil was still lined up with the corner is your dominant eye.

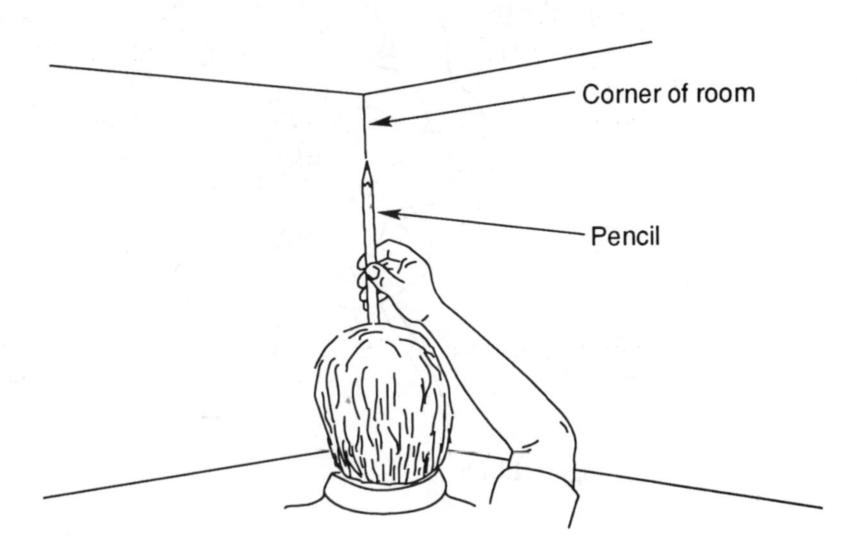

Figure 10.1.4 Determine your dominate eye

8. Select a spot in the center of the fold in the index card and stare at it with your dominant eye. (Cover the other eye with your hand.)

9. Stare at the card until you see it in the position shown in Figure 10.1.5. When the card changes position, move your head slowly from side to side.

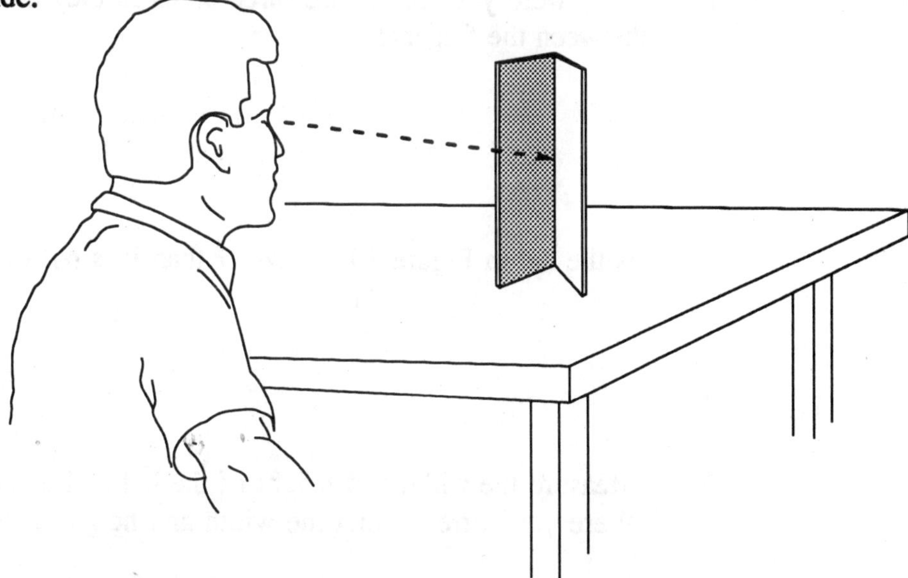

Figure 10.1.5 Stare at the index card until it appears to change position as shown

Part IV

10. Color the petals of Figure 10.1.6 yellow and the center red.

Figure 10.1.6

11. Stare at the flower for thirty seconds then stare at a blank, white sheet of paper.

RESULTS AND DISCUSSION

1. Why were you able to see three coins as they are rubbed quickly between the fingers?

2. Is the hat in Figure 10.1.2 wider than it is tall or taller than it is wide?

3. Measure the width and height of the hat and enter the result here. Were you correct about the width and height of the hat?

4. In what position did you see the card after staring at it for several seconds? This type phenomenon is a called an ________ illusion.

5. Did you continue to see the flower even when you looked at the blank sheet of paper?

6. Name some other, more common, optical illusions.

7. Do you think agricultural animals also see optical illusions? How do you think they might respond to optical illusions or other unusual things they see?

8. Do you think unusual sounds might affect animals in the same way as unusual things they see? How do you think animals react to unusual sounds?

NAME:______________________ DATE:____________ CLASS:____________

Chapter 10: Animal Behavior

Laboratory Exercise 10.2 The Behavior of Pill Bugs

BACKGROUND

Livestock producers have learned to use many of the behaviors of animals to increase production and to make handling easier while making the animal's life more comfortable. Ethology is the study of animal actions. When observations are made, the scientist sets out to find a practical use for the data.

Isopods (pill bugs and sow bugs to most of us) are harmless to humans and for the most part harmless to plants. They feed mostly on decaying plant material, although they sometimes feed on the live roots of potted plants, such as house plants sitting on the deck or in the greenhouse. In large numbers they may cause noticeable damage.

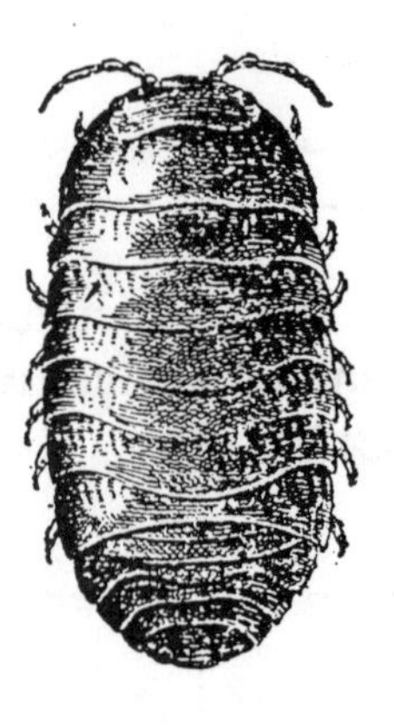

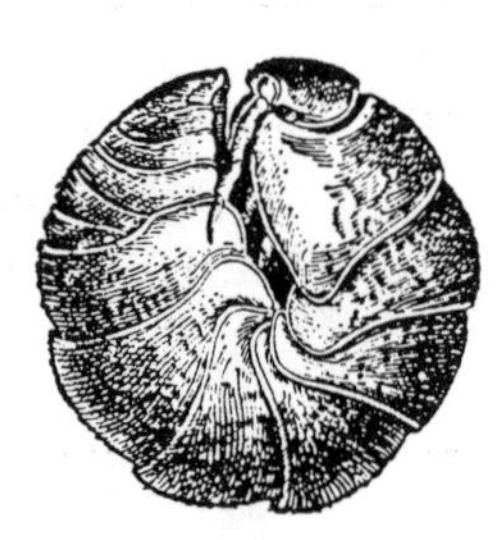

Figure 10.2.1 Pill Bug (USDA)

In this exercise you will use isopods to simulate the study of other animals. You will be asked to think about how the information you collect can be used.

For this exercise we will assume that isopods are dastardly pests. Your job is to study one specific aspect of an isopod's behavior to determine if it can be used in its control.

The Pill Bug

Pill bugs are also called sow bugs, wood lice, and rollie-pollies. They are neither sows, lice, or bugs. Pill bugs have seven pairs of legs, so they are not insects either. Scientists have named them isopods (equal legs). Isopods are in the class Crustacea. The pill bug is one of the few land-dwelling crustaceans. It is closely related to other crustaceans such as shrimp, crabs, and lobster. Seafood anyone?

OBJECTIVES

- To observe and record the behavior of an animal
- To interpret the results of animal behavior data
- To determine a practical use for animal behavior research data

EQUIPMENT

one piece of wood, approximately 4" x 12", any thickness
one three-inch block of wood

MATERIALS

three to ten isopods

SAFETY

There are no special safety precautions necessary for this exercise. Pill bugs will not harm humans.

PROCEDURES

1. Place the piece of wood on the block to form an incline plane at about a 35^0 - 45° angle.

2. Place the isopods in the center of the board one at a time.

3. Observe the direction of movement of each isopod and record the results in Table 10.2.1 in the Results and Discussion section.

4. The isopod may move erratically at first. You will have to make a judgment as to which is the most dominant direction. Record the dominant movement of each isopod by making a tally mark for each one in the appropriate box.

5. Observe in which direction each isopod is moving and record the results in Table 10.2.1.

    ↑ Up the incline

    ↓ Down the incline

← To the left

→ To the right

↗ or ↖ Up at a diagonal

↘ or ↙ Down at a diagonal

6. Repeat steps 2-5 for each isopod.

RESULTS AND DISCUSSION

1. Indicate in Table 10.2.1 the direction of movement of each isopod by making a tally mark in the appropriate cell.

Table 10.2.1 Movement of Isopods

Indicate the direction of movement of the isopods by making a mark in the appropriate cell. Make one mark for each isopod.

Isopod Number	Up ↑	Down ↓	Right →	Left ←	Down Diagonally ↘ or ↙	Up Diagonally ↗ or ↖
1						
2						
3						
4						
5						
6						
7						
8						
9						
10						

2. In which direction did most of the isopods move? Can you give a reason why this occurs?

3. Suppose the isopod were a particularly troublesome insect, could the information you have gathered be used to control it? Can you suggest a possible use?

4. Suppose hogs had the behavior you discovered about isopods in this exercise. How could we use that behavior in the production of swine?

NAME:______________________ DATE:______________ CLASS:______________

Chapter 10: Animal Behavior

Laboratory Exercise 10.3 Observing Animal Behavior

BACKGROUND

The study of animal behavior is called ethology. The behavior of domesticated animals is of interest to agricultural scientists because animals' behavior can be used to increase production, reduce handling costs, make life less stressful for the animals, and control pests.

Animal behavior is not an easy subject to study. The animals must be observed in detail, and conclusions are drawn from the data as to why the animals acted a certain way in a given situation and what implications that might have for animal care, handling, pest control, etc.

An ethogram is a comprehensive description of the behavior pattern of a species. The ethogram must be developed over a period of time with many observations. For example, a scientist might observe an animal each day at a different hour over a period of several months. The total observations can be used to define behavior patterns.

A Little Known Fact!

Suckling pigs have a particular order in which they line up to nurse at every feeding. This is an unusual behavior that is usually not known to the casual observer. The pigs line up at the sows teats in the order that they were born. You may have observed pigs wrestling for position when the sow lies down for feeding. Eventually they all settle down to a restful feeding time. The wrestling period is the method they use to find "their" teat. Ethologists discovered this behavior when they numbered the pigs in a litter as the pigs were born and made their way to the sow's teats. The scientists noticed that after the wrestling period at feeding time, the pigs were always lined up in numerical order!

In this exercise you will learn to collect data for an ethogram. Any animal easily observable in the classroom setting will be suitable. Your teacher will arrange for an animal to be observed. Hamsters, mice, dogs, and cats make convenient subjects. Farm animals nearby the school would be better, if available.

OBJECTIVES

- To collect data for an ethogram
- To define the behavior of the assigned animal
- To summarize animal behavior based on time
- To interpret ethogram data
- To make implications for the use of ethogram data

EQUIPMENT

clipboard

MATERIALS

notebook paper

SAFETY

For safety reasons and good research practice, do not disturb or interact with the animal in any way - only observe and record the behavior.

PROCEDURES

1. Study the sample ethogram data sheet in Table 10.3.1 and answer items 1-3 in the Results and Discussion section.

2. Your teacher will assign an animal to observe. Fill in the preliminary information on the ethogram data sheet in Table 10.3.2 in the Review and Discussion section.

3. Your teacher may assign or help you develop a research question based on the animal you are observing. For example, suppose you are observing a cow. The research might investigate what percent of time the cow spends grazing. See the sample data sheet.

4. Complete the ethogram key in Table 10.3.2 for the animal you are studying. See the data sheet prepared for observing cattle to get an idea for codes to use.

5. Sit or stand quietly at a distance, only close enough to clearly observe the animal.

6. Record every action the animal takes by coding in the action on the data sheet. You will be doing sequence sampling. That is, you will

record what behaviors the animal exhibits and in what sequence. Time of action is one factor being left out in this exercise.

RESULTS AND DISCUSSION

1. What conclusion might be drawn from the data collected in the sample ethogram data sheet? Is a problem evident?

2. Do you think that this problem would normally have been noticed by the casual observer? Why or why not?

3. What effect do you think the problem indicated in the sample ethogram data sheet might have on production, health, and comfort of the animal?

4. What was the most common activity you observed in the laboratory animal?

5. What conclusions can you draw from the data you collected?

6. Describe the practical use of ethology. Explain how data from ethograms could help livestock producers and handlers.

Table 10.3.1 Ethogram Data Sheet

Animal: Brood Cow, Angus, appears to be approximately 3 years old
Research Question: Are there any observable problems in this cattle herd?
Observer: Ima Sample, 3rd period Agri-science class
Date: September 23, 1993
Start Time: 10:00 A.M. End Time: 10:45 A.M.
Animal's Environment: Open pasture, excellent grazing
Weather conditions: 70ºF, Sunny

No.	Behavior	No.	Behavior	No.	Behavior
1	wlk	21	wlk	41	
2	grz	22	grz	42	
3	pst	23	pst	43	
4	grz	24	pst	44	
5	def	25	voc	45	
6	con	26	grz	46	
7	pst	27	wlk	47	
8	wlk	28	def	48	
9	grz	29		49	
10	pst	30		50	
11	slp	31		51	
12	pst	32		52	
13	voc	33		53	
14	pst	34		54	
15	grz	35		55	
16	wlk	36		56	
17	grz	37		57	
18	pst	38		58	
19	voc	39		59	
20	grz	40		60	

Key	Behavior	Key	Behavior	Key	Behavior
Grz	grazing	Voc	Vocalizing	Cud	chewing cud
Wlk	walking	Drk	Drinking	Run	running
Pst	distracted/annoyed by pests	Slp	no action, sleepy/tired	Def	defecating or urinating
Con	contact with another animal	Lay	lying down	Oth	Other

Table 10.3.2 Ethogram Data Sheet

Animal:____________________
Research Question:____________________
Observer:____________________
Date: ____________________
Start Time:__________ End Time:__________
Animal's Environment:____________________
Weather conditions: ____________________

No.	Behavior	No.	Behavior	No.	Behavior
1		21		41	
2		22		42	
3		23		43	
4		24		44	
5		25		45	
6		26		46	
7		27		47	
8		28		48	
9		29		49	
10		30		50	
11		31		51	
12		32		52	
13		33		53	
14		34		54	
15		35		55	
16		36		56	
17		37		57	
18		38		58	
19		39		59	
20		40		60	

Key	Behavior	Key	Behavior	Key	Behavior

NAME:____________________ DATE:__________ CLASS:__________

Chapter 10: Animal Behavior

Laboratory Exercise 10.4 The Pecking Order

BACKGROUND

This exercise will demonstrate one type of animal behavior - pecking order. This aggressive behavior was observed by a Norwegian scientist in his flock of chickens. One chicken was acting like a tyrant. Upon careful study, he found that there was a regular order of dominance established in the flock. Hen A would drive away all the other hens whenever she desired. Hen B would peck and drive away all hens except hen A. Hen C was dominant over all hens except Hens A and B. This order of dominance (pecking order) continued through the flock.

The pecking order was not obvious until the scientist recognized and studied his chickens as individuals. Thus he learned that "one of the most important techniques in the study of behavior in animals is the identification of individuals."

In this exercise you will observe social dominance of one fish over another. This same type of dominance is exhibited with most farm animals, including swine and cattle. Dairy cattle are known for their social order. Studies have shown that dairy cows enter the barn in a specific order every time - the dominant cow first and the least dominant last.

You will observe five sunfish (bluegill, or any pan fish or redear). The fish will be marked so that you can identify them. The fish should be put in a tank for at least two days prior to the observation so that a pattern of dominance will be established. You will find that they are very aggressive and that one fish will completely dominate the group.

OBJECTIVES

- To identify dominant behavior in a group of animals
- To list reasons dominant behavior contributes to natural selection

EQUIPMENT

one complete aquarium kit for the class
scissors

MATERIALS

paper towel
floating fish feed

SAFETY

Care should be taken when using scissors. Secure the fish in a paper towel and have a classmate hold it firmly. A sudden jump by the fish may cause you to injure yourself, the fish, or others.

PROCEDURES

1. Obtain five small bluegill fish (at least two inches long) from the teacher. These can be obtained from biological supply stores, or a local pet shop will be able to suggest a suitable substitute.

2. One at a time, wrap each fish in a moist paper towel with its tail exposed. Have one student hold the fish securely while another marks the fish. Mark the fish by cutting a different shape in the tips of their tail fins - <u>be sure to cut only the very tips</u>. Be assured that this is a very minor procedure and in no way hurts the fish or affects its swimming ability. Use Figure 10.4.1 and the following as a guide:

   Fish 1 - nip the top of the tail fin with the scissors
   Fish 2 - nip the bottom of the tail fin with the scissors
   Fish 3 - cut a "V" shape in the tail fin with the scissors
   Fish 4 - nip the top and bottom of the tail fin with scissors
   Fish 5 - no marking

Figure 10.4.1 Marking Fish for Identification

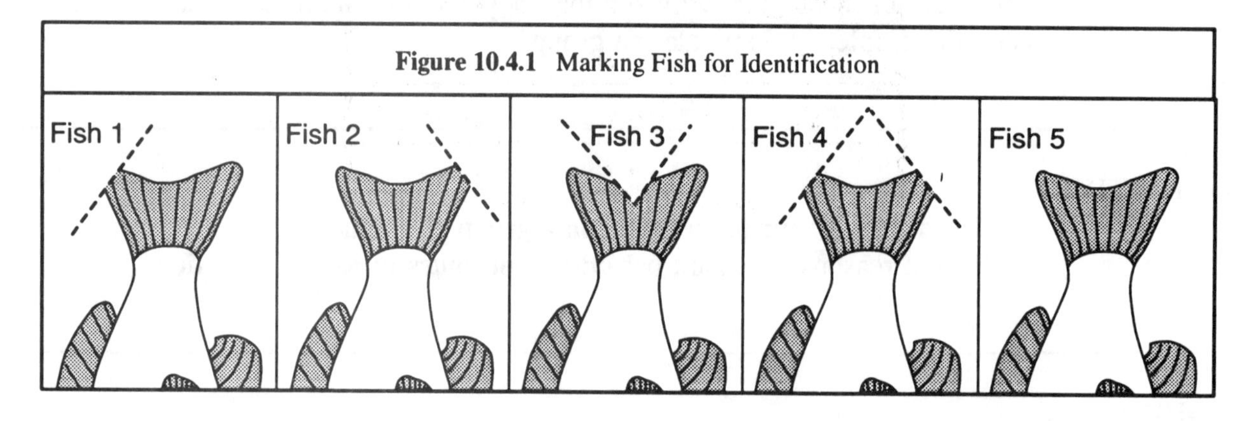

3. Do not feed the fish for twelve hours before the observation.

4. At the time of observation put a very small amount of feed on the surface to get the fish moving.

5. You may want to add more feed periodically during the observation. Complete the observation sheet included in the Results and Discussion section.

RESULTS AND DISCUSSION

1. Record each time you observe one fish act aggressively toward another.

Table 10.4.1

Dominance Observation Chart

Dominance Observation Chart
Fish ____ chased fish ____ away
Fish ____ chased fish ____ away
Fish ____ chased fish ____ away
Fish ____ chased fish ____ away
Fish ____ chased fish ____ away
Fish ____ chased fish ____ away
Fish ____ chased fish ____ away
Fish ____ chased fish ____ away
Fish ____ chased fish ____ away
Fish ____ chased fish ____ away
Fish ____ chased fish ____ away

2. Based on the data you collected, rank the fish by pecking order - most dominant first and least dominant last.

_________ most dominant

_________ second

_________ third

_________ fourth

_________ least dominant

3. Very often the dominance of one animal over another is a major problem in feeding livestock. How can a producer minimize the problem?

4. Besides less-dominant animals not getting enough to eat, what other problems might a very aggressive animal cause to the other animals?

5. What role do you think the establishment of a pecking order within groups might have played in natural selection?

NAME:______________________ DATE:____________ CLASS:____________

Chapter 11: Animal Genetics

Laboratory Exercise 11.1 Male/Female Ratio

BACKGROUND

The sex of an animal is determined by the matching of chromosomes from the mother and father and is determined at conception. Each body cell contains one pair of chromosomes called the sex chromosomes. Each gamete (sex cell from the parent) contains one half of the sex chromosome from the parent.

The female chromosome is referred to as *XX*. When the chromosome divides, one *X* chromosome is given to each gamete (egg). The male chromosome is referred to as *XY*, and when it divides, each gamete (sperm) receives an *X* or a *Y* chromosome but not both.

The two halves (in the sperm and the egg) are united at conception. If the zygote contains the *XX* sex chromosome, it will be female. If it contains the *XY* sex chromosome, it will be male. For this reason it is said that the male determines the sex of the offspring.

In this exercise you will demonstrate the probability of female and male offspring.

OBJECTIVE

- To demonstrate the probability of male and female offspring

---

EQUIPMENT

one cup

MATERIALS

two coins
masking tape

SAFETY

There are no special safety concerns in this laboratory exercise.

PROCEDURES

1. Wrap two coins with masking tape.

2. Mark one of the coins with a large *X* on both sides. This will represent the female gamete.

3. Mark the other coin with an *X* on one side and a *Y* on the other. This will represent the male gamete.

4. Place the two coins in the cup. Holding one hand over the mouth, shake the coins and toss them onto the table.

5. Record the results with a tally mark in Table 11.1.1 in the Results and Discussion section. There are only two possibilities - two *X*s or one *X* and one *Y*.

Figure 11.1.1 Coins representing female (XX) and male (XY)

6. Toss the coins nine more times and record the results in the "10 toss" row in the tally chart.

7. Toss the coins one hundred times and keep a record of the results by placing a tally mark in the appropriate cell of the "100 toss" row.

8. Determine the ratio of *XX* to *XY*.

9. Compile your results with those of your classmates.

RESULTS AND DISCUSSION

1. What is the probability of tossing an *XX*?

2. What is the probability of tossing an *XY*?

3. Record the results in Table 11.1.1.

Table 11.1.1 Sex Determination Tally Chart		
	XX	XY
10 tosses		
100 tosses		

4. What was the expected ratio of males to females?

5. What was the actual ratio of males to females after ten tosses?

6. What was the actual ratio of males to females after one hundred tosses?

7. Was the ratio after one hundred tosses more nearly the expected ratio than after ten tosses? Why?

NAME:__________________ DATE:__________ CLASS:__________

Chapter 11: Animal Genetics

Laboratory Exercise 11.2 Phenotype versus Genotype

BACKGROUND

Differences in animals can be described as differences in their genetic makeup and phenotype. Phenotype is the physical appearance of the animal, such as color, size, shape, and other characteristics.

An animal's phenotype is the result of genotype and environmental factors. Characteristics of individual animals are controlled by the animal's genes that were passed on by its parents. An animal gets half of its genetic makeup from each of its parents.

Every gene from the male is paired with a gene of the same type from the female. For example, the gene that controls the color of the animal's coat is made up of a pair of "coat color" genes, one from the father and one from the mother. A pair of genes that controls a specific characteristic is called an allele. If both of the genes are the same, that is, both genes call for a black coat or both call for a red coat, etc., the genes are said to be homozygous.

The offspring is said to be heterozygous if the genes are different. For example, an animal is heterozygous if one gene calls for a red coat color while the other calls for a black coat. The color of the offspring's coat will be determined by the dominant gene. This means that one gene will override the effect of the other gene.

Color coding is a good example of how genes transfer traits. The same general principles can be applied to other traits, such as horned or polled, tall or short, etc. In this exercise you will determine the probability of attaining various colors of hair coats when two heterozygous parents are mated.

OBJECTIVES

- To determine mathematically the expected color of offspring
- To compare the expected coat color with results obtained through scientific investigation

---

EQUIPMENT

two coins
cup
magic marker

MATERIALS

masking tape

SAFETY

There are no special safety concerns necessary with this laboratory exercise.

PROCEDURES

Part I

1. Complete the Punnett's square in item 1 in the Results and Discussion section before continuing this exercise. Assume that the uppercase *B* is dominant for black hair coat and that the lowercase *b* represents the recessive gene for red hair coat color.

Part II

2. Wrap two coins with masking tape.

3. Mark one side of both coins with a large *B*. This will represent black coat color. It is capitalized to represent black as the dominate color.

4. Mark the other sides of the coins with a lower case *b*. This will represent red coat color. In this case it is recessive, not dominant.

5. Place the two coins into the cup. Holding one hand over the mouth, shake the coins and toss them onto the table.

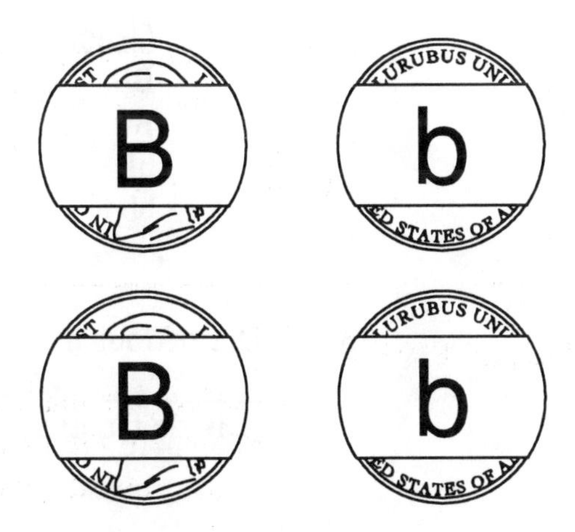

Figure 11.2.1 B represents black coat color and b represents red coat color

6. Record the results in Table 11.2.1 in the Results and Discussion section. There are only three possibilities - *BB*, *bb*, or *Bb*.

7. Toss the coins nine more times and record the results in the "10 toss" row in the tally chart.

8. Toss the coins one hundred times and keep a record of the results by placing a tally mark in the appropriate cell of "100 toss" row.

9. Determine the ratio of *BB*, *Bb*, and *bb*.

10. Compile your results with those of your classmates.

RESULTS AND DISCUSSION

1. Complete the Punnett's square below by placing the symbol for one gene from each parent in the appropriate cell.

	B	b
B		
b		

2. What color would each of the following offspring be if their genotype was

Genotype	Color
BB	________
Bb	________
bB	________
bb	________

3. Out of four offspring, how many would be expected to be black? ________. What percent of the offspring are expected to be black? (Divide the expected number by the total number, 4 in this case.) ____.

4. Out of four offspring, how many would be expected to be red? ________. What percent of the offspring are expected to be red? (Divide the expected number by the total number, 4 in this case.) ____.

5. Record the results in Table 11.2.1.

Table 11.2.1 Hair Coat Tally Chart			
	BB	Bb or bB	bb
10 tosses			
100 tosses			

6. What was the expected ratio of black to red?

7. What was the actual ratio of black to red after ten tosses?

8. What was the actual ratio of black to red after one hundred tosses?

9. Was the ratio after one hundred tosses more nearly the expected ratio of black to red than after ten tosses? Why?

10. After one hundred tosses how many offspring were homozygous?

11. After one hundred tosses, how many offspring were heterozygous?

NAME:____________________ DATE:____________ CLASS:____________

Chapter 12: The Scientific Selection of Agricultural Animals

Laboratory Exercise 12.1 The Gene Pool

BACKGROUND

The breeds of animals used in agriculture today have been developed through selective breeding over many hundreds of years. Selective breeding means that the parent animals were chosen to be bred together for a specific reason. Selective breeding is the opposite of chance breeding or breeding at random.

Wild animals practice random breeding. Natural selection plays a big part in controlling the gene pool in the wild. For example, the most productive breeds of chickens are white. In the wild, white chickens would probably not last very long. They would be much easier to see than darker colors that match the terrain.

Producers have good reasons to "selectively breed" their animals. There is an old saying that the way to get a faster race horse is to breed the two fastest horses to each other. Producers try to breed animals that will give them offspring with the trait they desire. Producers may be breeding for increased milk production, a certain color hair coat, strong legs, etc.

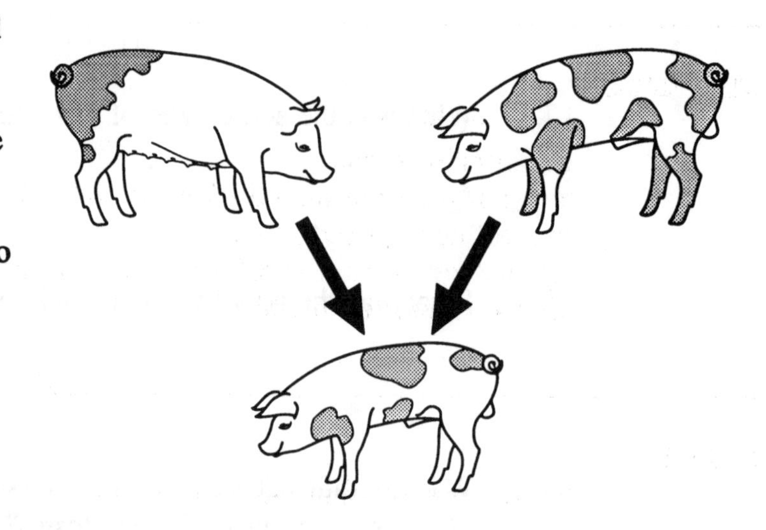

Figure 12.1.1 Selective breeding practice produce superior animals

Most animals from a particular breed all look alike and have approximately the same characteristics. Angus cattle, for example, are all black. Breeds such as Angus cattle probably got started when a person started selectively breeding the blackest cattle together. With each generation the gene pool for black was increased until today it is a very rare exception that a pure breed Angus cow is not all black.

In modern times, producers have learned to keep accurate records and to selectively breed animals with the best traits. In this exercise you will simulate the breeding of animals over several generations. The records you must keep concern the color of your animals.

OBJECTIVES

- To describe how animal characteristics are passed from generation to generation through genes
- To simulate the development of an animal with a pure color hair coat
- To distinguish between natural selection, random mating, and selective breeding

---

EQUIPMENT

one cup
watch

MATERIALS

ten white beans
ten speckled beans
a bag of mixed beans - 50% white and 50% speckled
ten toothpicks, dyed green
ten toothpicks, natural color or dyed brown
ten toothpicks, dyed white

SAFETY

There are no special safety precautions necessary with this exercise.

PROCEDURES

Part I

This exercise will simulate natural selection. The results can help explain why animals with certain traits have been able to survive while others became extinct or, as a group, lost certain characteristics. Nature helps control the gene pool of a population.

1. Mix the thirty toothpicks together - ten green, ten brown, and ten white.

2. Find a grassy area near the classroom as directed by the teacher. Each class member should have an area about ten foot square in which to work.

3. Throw the toothpicks high into the air above the grassy area for even distribution.

4. Now try to find as many toothpicks as possible in one minute.

5. Count the number of each color found. Record your answers in Table 12.1.1 in the Results and Discussion section.

Part II

This exercise will simulate random mating. That is, animals are bred without regard for their particular characteristics.

6. Without looking at the beans, place your hand in the 50/50 white and speckled bean sample supplied by your teacher and draw out twenty beans.

7. Count the number of beans of each color. Record the result as the initial genotype in Table 12.1.2.

8. Put the beans in your cup. These twenty beans represent your animal and its genotype for this exercise.

9. "Breed" your animal, in turn, to the animals of the students sitting in front of you, behind you to your left, and to your right. Do this by pouring your beans into their cup. Shake the beans to get a good mix.

10. Pour the beans into your hand. Close your hand, and without looking, count out twenty beans into your classmate's cup.

11. Record the new genotype in Table 12.1.2.

Part III

Selective breeding: You are to develop an animal with a solid white hair coat through selective breeding. A solid white coat will be represented by twenty white beans. The first person in the class to obtain a solid white coat wins.

12. Obtain ten each of white and speckled beans and a cup to hold them. This is the gene pool for your animal.

13. "Breed" your animal to a neighbor's animal by pouring your beans into his/her cup. Shake the beans to get a good mix.

14. Pour the beans into your hand. Close your hand, and without looking, count out twenty beans into your classmate's cup.

15. Record the new genotype on Table 12.1.3. This will simulate record keeping by the producer.

16. Continue to "breed" your animal to others in the class. You will naturally want to breed your animal to one that has more white color than yours. You may ask the classmate for his/her records and to examine his/her animal before committing to breed.

17. Conversely it is not in your interest to breed with an animal that has fewer white beans than yours.

18. The more breedings you have with animals better than yours, the better chance you have to obtain a pure white coat.

19. After a member of the class has reached the objective or time is called by the teacher, return the beans to the source and complete the worksheet.

RESULTS AND DISCUSSION

1. Using the data from Part I, record the number of each color toothpick you found and the total number found.

Table 12.1.1 Natural Selection			
Number of White	Number of Green	Number of Brown	Total Toothpicks Found

2. What does the data in Table 12.1.1 imply about the color of an animal and its ability to hide from its prey?

3. Under what conditions would you have found more of the other colors of toothpicks?

4. What results did you expect with the random mating of animals as simulated in Part II?

5. Record the results of the random mating simulated in Part II.

Table 12.1.2 Random Breeding

	White Beans	Speckled Beans
Initial genotype		
Genotype after breeding 1		
Breeding 2		
Breeding 3		
Breeding 4		

6. Were the results of the random breeding exercise what you expected? Explain.

7. Record the results of your selective breed exercise.

Table 12.1.3 Selective Breeding Record

	Number of White Beans	Number of Speckled Beans
Initial gene pool	10	10
Gene pool after breeding 1		
Breeding 2		
Breeding 3		
Breeding 4		
Breeding 5		
Breeding 6		
Breeding 7		
Breeding 8		
Breeding 9		
Breeding 10		
Breeding 11		
Breeding 12		
Breeding 13		

8. If you were not able to develop a pure animal in the above exercise, how many more breedings do you think it would take to develop a pure animal?

9. Suppose the animals in this exercise are cattle, and it takes you eighteen breedings to develop a pure white animal. How many years of development would this represent? It takes nine months for a cow to have a calf, and a cow must be one year of age before they can be breed.

10. Suppose the gene pool was much larger for the above exercise - maybe two hundred white and two hundred speckled. How many breedings would it take to develop a pure animal?

11. Suppose you wanted to start a new breed of animal. How would you go about it and how long do you think it would take?

12. Why is it important to keep records on animals?

13. Why is the exercise in Part III called selective breeding?

NAME:______________________ DATE:____________ CLASS:____________

Chapter 13: The Reproductive Process

Laboratory Exercise 13.1 Examining Sperm and Egg Cells

BACKGROUND

Understanding the reproductive process has always been important in the economic production of agricultural animals. The tremendous growth in the use of artificial insemination has made it more important than ever for producers as well as researchers to clearly understand the process.

Reproduction in animals is achieved by each of two parents contributing genetic material to the young. Each of the two parents creates reproductive cells called gametes. The male gamete is known as a sperm cell (Figure 13.1.1), and the female gamete is known as an egg cell.

The gametes are so small that they cannot be seen with the naked eye. In this exercise you will observe semen and ovules with a microscope. Activities in the first part of this exercise will use prepared slides. Part II of the exercise will give you a chance to examine live semen.

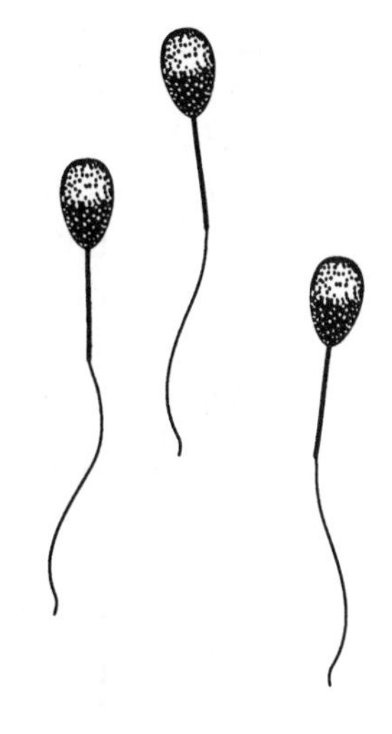

Figure 13.1.1 A sperm from the male fertilizes an egg from the female in the reproduction process

---

OBJECTIVES

- To compare the size and shape of sperm cells and egg cells
- To demonstrate the use of a compound microscope in observing sperm cells and egg cells

---

EQUIPMENT

one prepared slide of sperm
one prepared slide of an ovary
microscope with 10X, 40X and 100X (oil immersion) objectives

MATERIALS

immersion oil if a 100X objective is to be used

SAFETY

- Slides are made of glass so you should use care in handling them to prevent damage to the slide and injury to yourself.
- Never use direct sunlight as a light source for a microscope. The reflection of sunlight off the mirror and through the microscope could damage your eye.
- To avoid eye strain, keep both eyes open while looking through a microscope.

PROCEDURES

1. Switch the microscope to low power (10X) by turning the nosepiece until it clicks into place. The eyepiece magnifies objects ten times and the low-power objective magnifies the eyepiece ten times for a total magnification of one hundred times.

2. If your microscope has an electric light source, plug in the cord and turn on the light. If it has a mirror, turn the mirror toward a light source such as a lamp or open window. **Caution - Never use direct sunlight as a light source. Damage to your eye could result.** Adjust the mirror until the field of view is fully illuminated.

3. Place the prepared slide of sperm on the microscope stage. Hold the slide in position by placing the slide under the stage clips.

4. While looking at the low-power objective **from the side**, as shown in Figure 13.1.2, use the coarse adjustment to lower the objective as close as possible to the stage without touching it.

Figure 13.1.2 View the objective from the side while using the coarse adjustment

5. Look through the eyepiece and turn the coarse adjustment slowly to move the objective **away** from the stage. Continue to turn until the slide comes into focus.

6. Microscope lenses and slides are very fragile. Never lower the objective while looking through the eyepiece.

7. Locate sperm cells by adjusting the slide and using the fine adjustment to bring the sperm into sharp focus.

8. If your microscope has a 100X (oil immersion) objective, ask your teacher for further instructions on the use of immersion oil before continuing.

9. While looking at the objectives from the side, switch to high power by turning the nosepiece until it clicks into position. Focus with the fine adjustment.

10. Sketch a sperm cell in the space provided in item 3 of the Results and Discussion section.

11. Switch back to low power. Replace the sperm slide with a prepared slide of an ovary.

12. Repeat steps 4 and 5 above to focus on the ovary using low power. Use the fine adjustment as necessary. Sketch the ovary in the space provided in item 4 of the Results and Discussion section.

13. Notice the many round maturing follicles that are swollen with fluid. Note the varying sizes of the eggs within the follicles.

14. Locate a mature follicle and switch to high power. Use the fine adjustment to focus. Sketch the mature follicle in the space provided in item 5 of the Results and Discussion section.

RESULTS AND DISCUSSION

1. What kind of light source does your microscope have?

2. Complete Table 13.1.1 on the magnification of each objective and eyepiece of your microscope. The total magnification is determined by multiplying the magnification of the eyepiece (10X) by the magnification of the objective.

Table 13.1.1 Microscope Magnification			
Objective	Magnification of Objective	Magnification of Eyepiece	Total Magnification
Low Power			
High Power			
Other (if available on your microscope)			

3. Make a detailed drawing of sperm under low power and high power. It may be helpful to you to review Figure 13-10 on page 210 in the textbook *The Science of Animal Agriculture*.

Sperm under low power	Sperm under high power

4. Make a detailed drawing of the ovary under low power. It may be helpful to you to review Figure 13-11 on page 210 in the textbook *The Science of Animal Agriculture*.

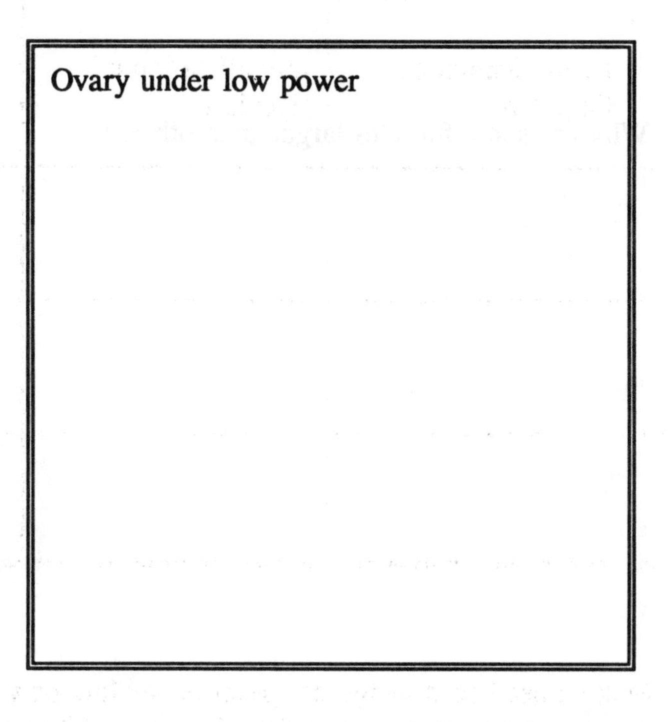

5. Make a detailed drawing of a mature follicle and ovum (egg).

Mature follicle and ovum under high power

6. Describe the shape of a sperm. How does the shape of a sperm help it perform its function?

7. Why are some follicles larger than others?

NAME:________________________ DATE:______________ CLASS:______________

Chapter 13: The Reproductive Process

Laboratory Exercise 13.2 Evaluating Bull Semen

BACKGROUND

Semen for artificial insemination is frozen and stored in liquid nitrogen tanks at -320°F. Semen from bulls, stallions, and rams can be frozen, stored, and thawed successfully. Semen from boars is usually used fresh because of problems with sperm livability.

Once the semen is collected, it is examined in the laboratory under a microscope. It is checked for foreign material and for quality. Quality is determined by the number of sperm in a milliliter of semen, how active the sperm are, and the shape of the sperm. Very active sperm are desirable because of the distance they must travel to reach the oviduct of the female.

The motility of sperm is defined as the rate of movement. Motility is one measure of the fertility of the sperm. Generally the greater the number of motile spermatozoa, the more likely it is that conception will take place. An estimate of motility is made while examining the semen under a microscope. Motility is only a rough estimate of semen quality. Also, a single test may have limited value since the quality of semen samples from the same bull varies over a period of time.

Semen from some bulls may be of sufficient quality for natural breeding, but insufficient quality to be frozen and thawed in an artificial insemination program. This laboratory procedure is a sample of the tests used by artificial insemination companies to evaluate semen quality.

Bull semen will be used in this laboratory exercise. You will be thawing and preparing semen for this exercise just as you would if you were going to artificially inseminate an animal.

OBJECTIVES

- To evaluate the motility of bull semen
- To demonstrate the use of a compound microscope in observing sperm motility
- To explain how to evaluate a semen sample for motility
- To explain how to prepare a straw of bull semen for insemination

---

EQUIPMENT

compound microscope
laboratory thermometer
widemouthed thermos
hot plate
pan
two slides
one coverslip
eye dropper
scissors

MATERIALS

straw of bull semen
paper towel

SAFETY

- Liquid nitrogen can cause a frostbite-like injury if it contacts the skin. Follow carefully the precautions in this exercise and the instructions of your teacher.
- Slides are made of glass so you should use care in handling them to prevent damage to the slide and injury to yourself.
- Never use direct sunlight as a light source for a microscope. The reflection of sunlight off the mirror and through the microscope could damage your eye.
- To avoid eye strain, keep both eyes open while looking through a microscope.

PROCEDURES

1. Switch to low power (10X) on the microscope by turning the nosepiece until it clicks into place.

2. If your microscope has an electric light source, plug in the cord and turn on the light. If it has a mirror, turn the mirror toward a light source such as a lamp or open window. **Caution - Never use direct sunlight as a light source. Damage to your eye could result.** Adjust the mirror until the field of view is fully illuminated.

3. Prepare a container of warm water to between 94° and 100°F. You may need the hot plate and pan to heat the water. Place the thermometer and warm water in a thermos and adjust until the temperature is at the desired level.

4. Obtain a straw of semen from the teacher. The teacher will demonstrate removal of the straw from the semen tank.

5. Place the straw in the water bath for at least one minute but no longer than fifteen minutes.

6. Warm two slides and one coverslip to approximately 100°F. Prepare a bowl of warm water - about like the temperature of bath water. Put the slides and coverslip in the water. Five minutes or more may be needed for warming.

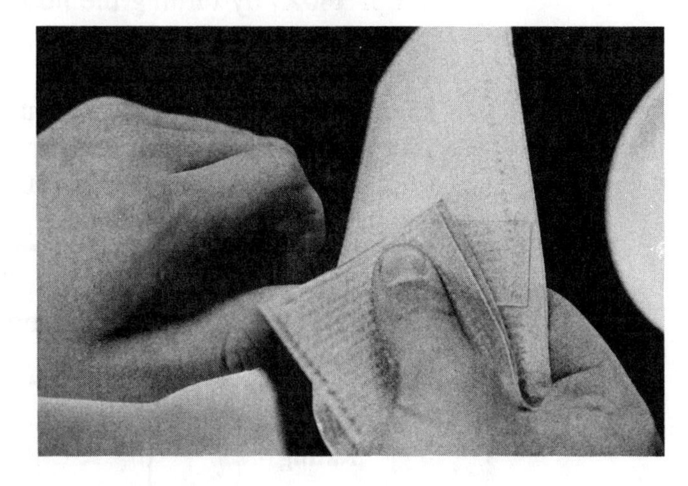

Figure 13.2.1 Dry the slide quickly after removing it from warm water

7. A slide at 100°F is barely above human body temperature. Use one slide to test for temperature. Take it out of the water and quickly dry it with a paper towel, Figure 13.2.1. Touch the slide to the inside of the wrist (like testing baby formula) and press down gently. **Caution - the slide is glass and will cut you.** If the slide feels cool, it needs to be warmer; if it feels much warmer than the skin, it should be cooled by moving the plate further away from the light.

8. Remove the thawed straw from the bath water and mix it by inverting it two to three times.

9. Follow the directions of your teacher to open the straw. Place one drop of semen on the warm, clean slide (not the one you used for testing). Place a coverslip over the semen.

10. Place the slide of sperm on the microscope stage. Hold the slide in position by placing the slide under the stage clips.

11. While looking at the low-power objective **from the side**, as shown in Figure 13.1.2, use the coarse adjustment to lower the objective as close as possible to the slide without touching it.

12. Look through the eyepiece and turn the coarse adjustment slowly to move the objective **away** from the stage. Continue to turn until the slide comes into focus.

13. Microscope lenses and slides are very fragile. Never lower the objective while looking through the eyepiece.

14. While looking at the objectives from the side, switch to high power (40X) by turning the nosepiece until it clicks into position. Focus with the fine adjustment.

15. Observe individual sperm cells and their movement. Strong healthy sperm will move in a straight line across the viewing field. Rate the sample with the scale found in Table 13.2.1.

Table 13.2.1 Sperm Motility Rating Scale

Rating	Criterion
Excellent	80% or more of the sperm are moving vigorously.
Very Good	70 to 80% of the sperm are moving vigorously.
Good	50 to 70% motility
Poor	30 to 50% motility
Very Poor	Less than 30% motility

16. Obtain a second sample of semen from the teacher. The teacher will have prepared the second sample with a treatment unknown to the students. Follow steps 7 through 15 to score the second sample on motility.

RESULTS AND DISCUSSION

1. Explain why a company that specializes in artificial insemination would want to check sperm motility on a regular basis.

2. Why does semen from a particular bull need to be checked over a period of time?

3. Record the motility scores of the two semen samples observed in this laboratory exercise in Table 13.2.2.

Table 13.2.2 Sperm Motility Rating

	% Motile Sperm Cells	Rating
Sample 1		
Sample 2		

4. What could have accounted for the difference, if any, in motility rating of the semen samples?

5. Why are vigorous, quick moving sperm desirable?

6. What might be the result if the motility of semen is checked using a cold slide sample? What might be the result if the slide is too hot?

7. Why are more-motile sperm considered more fertile?

8. This semen motility test is called a subjective measure. What does this mean? Why might this be a problem?

NAME:________________________ DATE:__________ CLASS:____________

Chapter 14: Animal Growth and Development

Laboratory Exercise 14.1 Examining Animal Cells

BACKGROUND

Cells are the building blocks of plant and animal life. An animal's body contains a number of different kinds of cells. Cells are the fundamental organizational unit of life.

There are fat cells, muscle cells, sex cells, etc. Each type of cell performs a particular function in the body. During both pre- and postnatal growth the increase in size is a result of the cells increasing in size or in number.

In this laboratory exercise, you will examine an animal cell. You will collect cells from your cheek to examine as a sample of animal cells.

OBJECTIVES

- To observe animal cells
- To identify the major parts of an animal cell
- To demonstrate the use of stain to observe cell parts

EQUIPMENT

compound microscope
one microscope slide
one coverslip
dropper

MATERIALS

one toothpick
methylene blue stain
paper towel

SAFETY

Handle the methylene blue stain with care. It will stain clothes and skin. You may want to wear a laboratory apron during this exercise.

PROCEDURES

1. Place a slide on a paper towel and put a drop of water in the center of the slide.

2. Gently scrape the inside of your cheek with the blunt end of a toothpick. Only slight pressure is needed.

3. The toothpick will have a tiny drop of saliva, which contains several cells. The cells are too small for you to see with the naked eye. Stir the toothpick into the drop of water on your slide, then put the toothpick in the trash can.

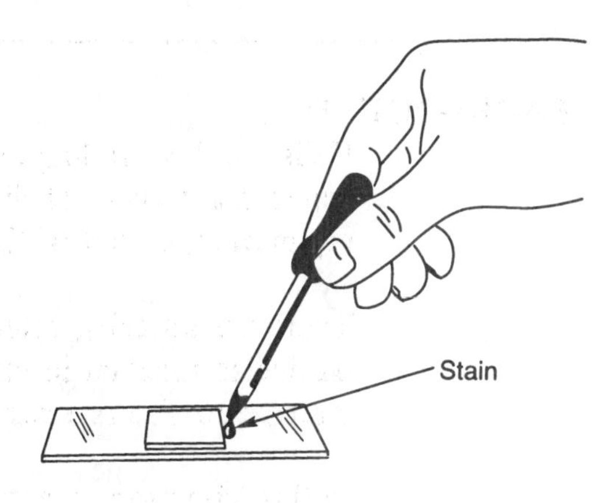

Figure 14.1.1 Place a drop of stain on one side of the cover slip

4. Gently place a coverslip over the drop on the slide. Place a drop of methylene blue or similar stain on the slide next to the edge of the coverslip, as shown in Figure 14.1.1. Fold a paper towel as shown in Figure 14.1.2 and touch it to the moisture on the opposite side of the coverslip from the stain. This will draw the stain under the cover slip.

5. Switch the microscope to low power (10X) by turning the nosepiece until it clicks into place. The eyepiece magnifies the object ten times, and the low-power objective magnifies the eyepiece ten times for a total magnification of one hundred times.

6. If your microscope has an electric light source, plug in the cord and turn on the light. If it has a mirror, turn the mirror toward a light source such as a lamp or open window. **Caution - Never use direct sunlight as a light source. Damage to your eye could result.** Adjust the mirror until the field of view is fully illuminated.

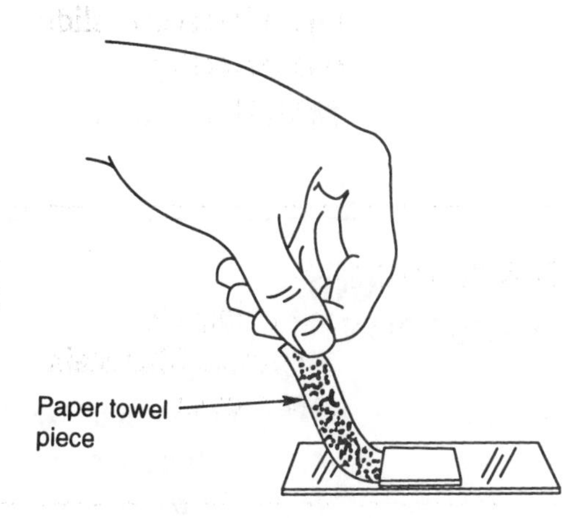

Figure 14.1.2 Use a paper towel to draw the stain under the cover slip

7. Place the slide of cheek cells on the microscope stage. Hold the slide in position by placing the slide under the stage clips.

8. While looking at the low-power objective **from the side**, as shown in Figure 14.1.3, use the coarse adjustment to lower the objective as close as possible to the stage without touching the slide.

Figure 14.1.3 View the objective from the side while using the coarse adjustment

9. Look through the eyepiece and turn the coarse adjustment slowly to move the objective **away** from the stage. Continue to turn until the slide comes into focus.

10. Microscope lenses and slides are very fragile. Never lower the objective while looking through the eyepiece.

11. You will see clumps of cells. Locate one or more cells that are spaced enough for you to see clearly by adjusting the slide and using the fine adjustment to bring them into sharp focus.

12. While looking at the objectives from the side, switch to high power (40X) by turning the nosepiece until it clicks into position. Focus with the fine adjustment. **Do not adjust the coarse focus.** Adjust the light as necessary to best observe the cell and its parts.

13. Complete each item in the Results and Discussion section.

14. Dispose of the slide as directed by your teacher.

RESULTS AND DISCUSSION

1. Draw a cheek cell as observed under high power in the space provided. Locate and label the cell membrane, nucleus, and cytoplasm. The definitions of these cell parts are given here to assist you in locating them.

Cell membrane: the outer covering of the cell.

Cytoplasm: the living substance within a cell excluding the nucleus.

Nucleus: the control center of the cell; usually the largest and most conspicuous structure within the cell.

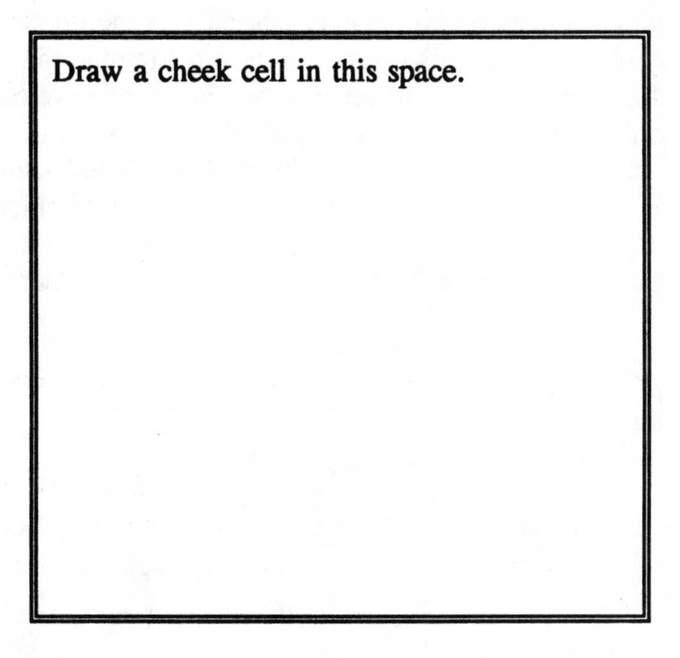

2. Why was blue stain used on the slide?

3. How many cells do you estimate you have on the slide?

NAME:______________________ DATE:______________ CLASS:______________

Chapter 14: Animal Growth and Development

Laboratory Exercise 14.2 Examining a Chick Embryo

BACKGROUND

Fertile chicken eggs require twenty-one days at the proper temperature and humidity to hatch. In this exercise you will examine the developing embryo at twenty-four, forty-eight, and seventy-two hours of development. You will examine the rapid development of a chick embryo. It is difficult to study the embryo development of most agricultural animals because they develop in the placenta within the mother's body. Embryos such as chicks, which develop outside the body, provide an easy means of studying embryo development. Also, chicken embryos are very much like those of other mammals.

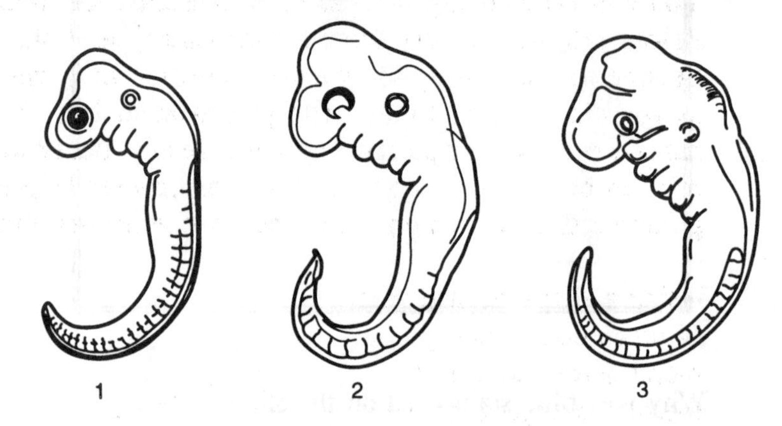

Figure 14.2.1 Embryos at early stages of development closely resemble one another: (1) fish, (2) chick, (3) human

OBJECTIVES

- To identify characteristics of twenty-four, forty-eight, and seventy-two hour old embryos
- To identify body structures of developing embryos

EQUIPMENT

sharp pointed dissecting scissors
magnifying glass or dissecting microscope
petri dish
dropper (optional)

MATERIALS

one fertile egg incubated for twenty-four hours
one fertile egg incubated for forty-eight hours
one fertile egg incubated for seventy-two hours
paper towels

SAFETY

Scissors should be used carefully as directed in the procedures section or by the teacher.

PROCEDURES

1. You will need an egg that has been incubated for twenty-four hours. Before selecting the egg from the incubator, mark the egg on the top. Be sure to keep the egg in the same position as it was in the incubator as you pick it up and move it to your workstation. Nest the egg in a paper towel in your petri dish. Let it remain undisturbed for two minutes before you continue. These procedures help ensure that the germinal disk will be on top of the yoke when you open the egg.

2. Using a sharp pointed pair of dissection scissors, pierce a tiny hole in the top of the egg. Slowly cut away a circular shape in the top of the egg, as shown in Figure 14.2.2. Remove any pieces of shell.

Figure 14.2.2 Opening an egg

3. Use a dropper to remove as much albumen as possible. This will make observation of the embryo easier. The embryo should be in sight on top of the egg. You may have to gently rotate the yoke with the dropper to locate the embryo.

4. Observe the embryo under a dissection microscope or hand lens and prepare a sketch in the Results and Discussion section.

5. Locate and label the head, heart, brain, eye, and somites (which will develop into the backbone).

6. Measure the embryo and record your answer.

7. Repeat steps 1-6 as directed by your teacher to examine eggs that have been incubated forty-eight and seventy-two hours. Use Figures 14.2.3 through 14.2.5 to locate developing parts of the embryo.

8. Dispose of your eggs as directed by your teacher.

RESULTS AND DISCUSSION

1. Describe the appearance of each of the parts listed below and sketch their appearance in the egg in the space provided.

   a. head ______________________________

   b. eye ______________________________

   c. heart ______________________________

   d. somites ______________________________

   e. brain ______________________________

2. How many pairs of somites can be observed?

3. Where do the blood vessels lead to and what function do you think they perform?

4. How is the germinal disk held on top of the egg?

5. If eggs of varying stages of development are opened, list the changes noted with each subsequent stage. Instead of the above exercise, your teacher may instruct you to complete the following exercise using the drawings in Figures 14.2.3 to 14.2.5.

Twenty-four hour chick embryo

____

____

____

____

Forty-eight hour chick embryo

____

____

____

____

Seventy-two hour chick embryo

____

____

____

____

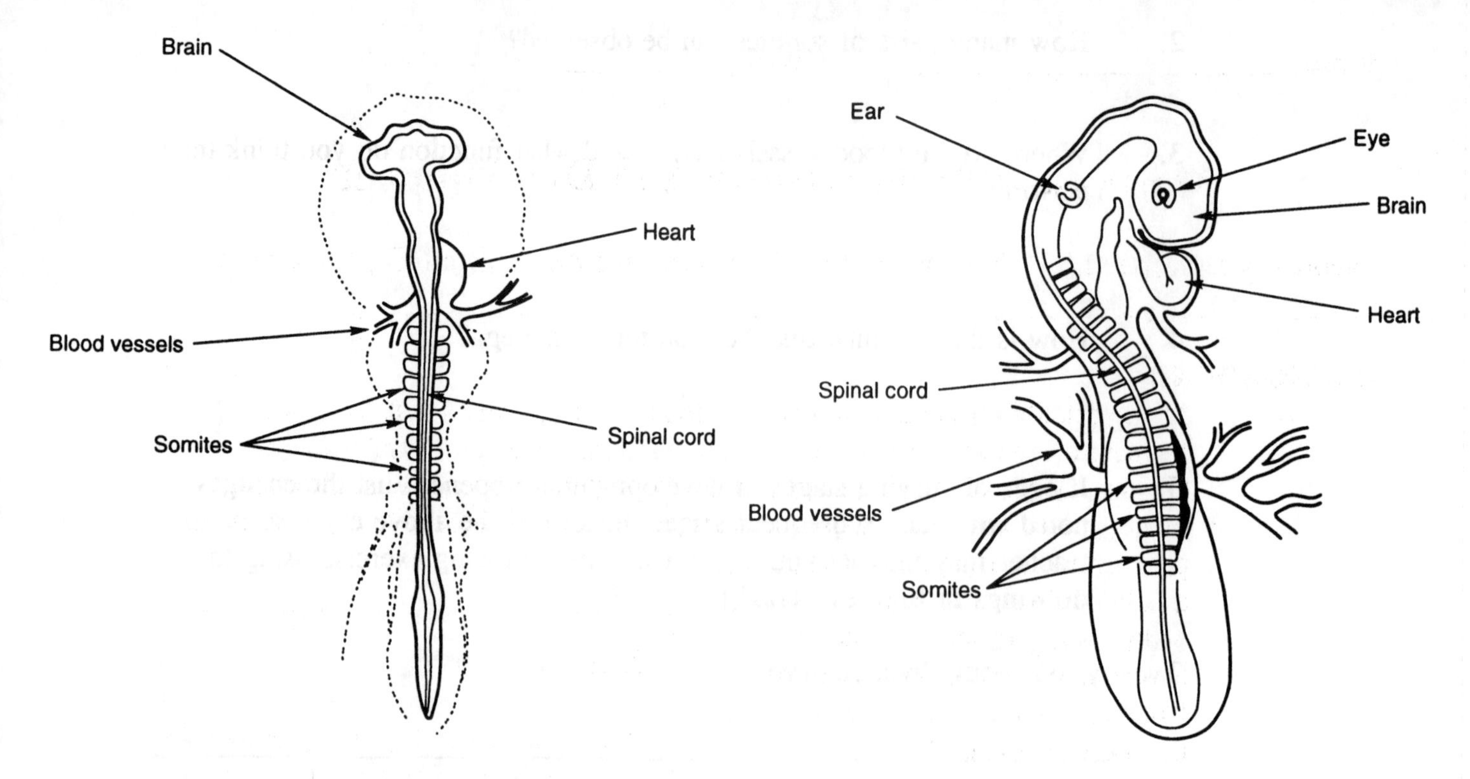

Figure 14.2.3 24 hour chick embryo

Figure 14.2.4 48 hour chick embryo

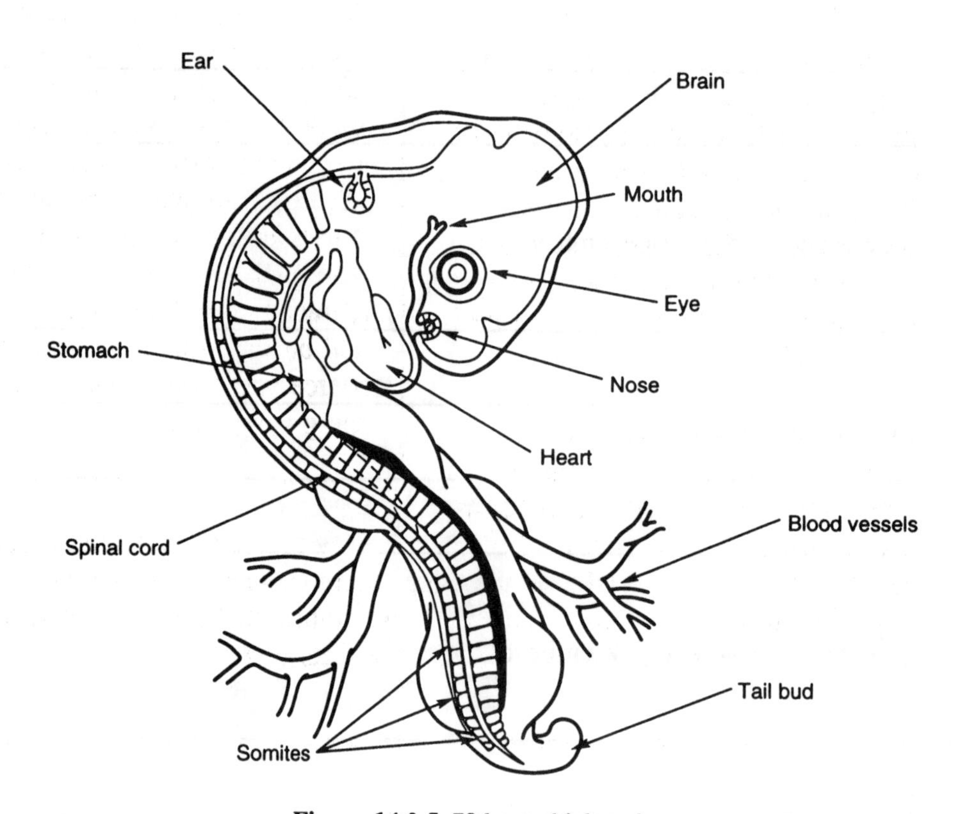

Figure 14.2.5 72 hour chick embryo

NAME:______________________ DATE:____________ CLASS:____________

Chapter 14: Animal Growth and Development

Laboratory Exercise 14.3 The Effect of Temperature on Animal Development

BACKGROUND

Most agricultural animals are warm blooded. This means that their body temperature must be kept at a constant temperature to grow and remain healthy. If the air temperature is not kept within a reasonably comfortable range, it will have a number of effects on the animal, most of which result in poor growth. If an animal is kept in a cold environment, it must use part of its energy to maintain body temperature. This energy comes from the animal's feed. When feed is used to maintain body temperature, it cannot be used for growth of the animal.

If temperatures are too hot, animals may go off feed. They appear listless and lie or stand resting most of the time. The less they move, the less heat their muscles generate. If they eat less, their bodies give off less heat. The animal tries to stay cool by moving less, staying in the shade, and eating less.

In this experiment you will use insects to observe the effects of temperature on animals. Insects are neither warm-blooded nor cold-blooded. They are classified as poikilothermic. Their bodies generally take on the temperature of their environment. Although not affected in exactly the same way as warm-blooded farm animals, insects are easy to use in a demonstration of the effects of temperature.

> **Energy Sources**
>
> Almost all energy on earth comes from the sun. The exceptions are nuclear and gravitational energy. When an animal eats plants, the energy the plant stored from sunlight is released into the animal's body. That energy warms the body as energy is released. Animals that eat meat get energy in an indirect way from plants. If the energy source is traced down the food chain it will be found that the original source was from plants that store energy from the sun.

OBJECTIVES

- To determine the relationship between temperature and the development of an animal
- To observe the four stages of development of the mealworm

EQUIPMENT

four small jars with lids - baby food jars work well or insect vials may be substituted
one nail
hammer
one beaker
four thermometers
magnifying glass
straight pin

MATERIALS

masking tape
one sheet of paper
mealworm colony with mealworms in all stages of development (also called golden grubs)

SAFETY

- Safety glasses should be worn by the user and observers if a hammer and nail are used to punch holes in lids.
- Mealworms and the adults are harmless to humans.

PROCEDURES

1. Obtain a sample population of mealworms from your teacher and place them in a beaker. Select from the sample all the pupae - see Figure 14.3.1. Obtain another sample if you do not have eight or more pupae. Return the balance of the sample population to the teacher.

2. Place the pupae on a sheet of paper. Examine each with a magnifying glass. Estimate their stage of development and select four

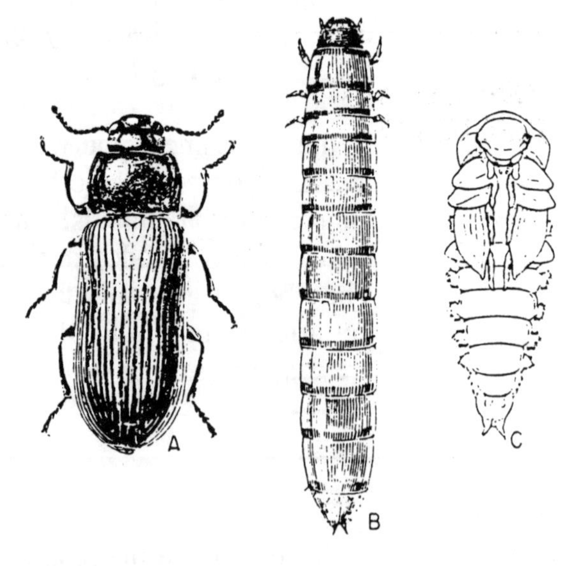

Figure 14.3.1 Stages of development of the mealworm. California Agricultural Experiment Station Bulletin.

that appear to be in about the same stage of development. Return all but the four selected pupae to the teacher.

3. Place each pupa in a separate jar or vial.

4. Punch air holes in the lids of the jars. If vials are used, cover the mouth of the vial securely with masking tape. Punch several tiny air holes in the tape with a straight pin.

5. Select four locations of varying temperature levels to store your pupae for development. Suggestions include the refrigerator, a garage, in the classroom at air temperature, in a basement, or outdoors. Be sure to keep the containers out of direct sunlight. Your teacher will probably want to arrange to store containers from all class members together.

6. Mark each container with a piece of masking tape. Write on the tape the place at which the pupae will be stored.

7. Place a thermometer beside the containers at each location. Take daily readings until adults emerge from the pupae. Record the temperatures in Table 14.3.1. After the pupa hatches, determine the average temperature and record it in Table 14.3.1.

8. Record the number of days it took the adult to emerge at each location in Table 14.3.2. Also record the average temperature at each location.

9. More accurate results can be obtained if an average of the class is taken. Compile the class data and enter it in Table 14.3.2.

RESULTS AND DISCUSSION

1. Fill in the location where each pupa is stored during development. Record the temperature each day at each location until the pupae hatch. Average the temperature for each location after the pupa hatches.

Table 14.3.1 Record of Temperature at Each Location

# 1 Location:		# 2 Location:		#3 Location:		#4 Location:	
Day	Temp	Day	Temp	Day	Temp	Day	Temp
1		1		1		1	
2		2		2		2	
3		3		3		3	
4		4		4		4	
5		5		5		5	
6		6		6		6	
7		7		7		7	
8		8		8		8	
9		9		9		9	
10		10		10		10	
11		11		11		11	
12		12		12		12	
13		13		13		13	
14		14		14		14	
15		15		15		15	
Average Temp.		Average Temp.		Average Temp.		Average Temp.	

2. Record the number of days it took for your pupae to emerge into an adult at each location in Table 14.3.2. List location and average temperature from Table 14.3.1. In the last column, record the class average of number of days to development.

Table 14.3.2 Mealworm Pupae Development					
Location	Average Temperature	Starting Date	Ending Date	Number of Days to Emerge	Class Average - Number of Days to Emerge

3. Why is the class average more accurate than your data alone?

4. Explain how temperature affects the development of mealworms.

NAME:______________________ DATE:__________ CLASS:____________

Chapter 15: Animal Nutrition

Laboratory Exercise 15.1 Analyzing Feeds for Nutrient Content

---

BACKGROUND

Animals must have nutrients in each of six major classes. These are water, proteins, carbohydrates, fats, vitamins, and minerals. Each of these classes of nutrients serves a specific function in the metabolism of the animal. The term *metabolism* refers to all the chemical and physical processes that take place in the animal's body. These processes provide energy for all of the animal's activities and body functions, such as maintenance of the body, growth, tissue repair, and the breakdown of food within the digestive system.

One type of feed may supply several of the needed nutrients, but usually a certain feedstuff contains a concentration of a particular nutrient. A feedstuff is generally a feed component that producers would not normally feed by itself, but combined with other types of feedstuff it comprises a complete diet for the animal.

In this laboratory exercise you will learn to analyze feeds for the presence of three organic compounds found in plants and animals. Chemicals called indicators will be used to determine the presence of protein and carbohydrates (sugars and starch). A color change of an indicator is usually a positive test for the presence of an organic compound. A physical test will be made to determine the presence of fat.

OBJECTIVES

- To identify feeds that contain carbohydrates, protein, and/or fat
- To demonstrate the use of chemical tests to indicate the presence of nutrients
- To determine by physical means the presence of fat

EQUIPMENT

six test tubes with test tube rack
mortar and pestle
test tube holder
four droppers
safety goggles
hot plate

one 250-ml beaker
four 50-ml beakers

MATERIALS

feed used for animals or humans*
- slice of white potato
- wheat kernels
- peanuts or soybeans
- protein supplement pellets
- corn

masking tape
brown wrapping paper or brown paper bag
iodine solution (with dropper)
Benedict's solution (with dropper)
Biuret solution (with dropper)
distilled water

* Any material that might be used as feed for animals may be substituted. Some feed may need to be cooked to soften it for grinding.

SAFETY

- Wear safety goggles at all times during this exercise.
- If any chemicals are spilled, rinse it from your skin with water and call your teacher immediately.
- Handle the hot plate and heated test tubes with care. Use the test tube holder as directed.

PROCEDURES

Part I Test for Starch

1. Prepare the first sample by grinding a slice of potato using the mortar and pestle. Continue grinding until it is completely ground. Add a small amount of distilled water - enough to get the entire mixture into a waterlike solution. Grind until the feed and water are thoroughly mixed and the solids are small enough to pass through a dropper.

2. Pour this mixture into one of the small beakers. Use masking tape to label the beaker "#1 Potato."

3. Wash the mortar and pestle and repeat steps 1 and 2 to prepare each of the four other feed materials. Label each beaker appropriately.

4. Use masking tape to label five test tubes with the number and names of the feed preparation. Label a sixth test tube as "water."

Figure 15.1.1 Prepare samples of various foods

5. Using a clean dropper for each sample, place ten drops of each feed preparation into the appropriately labeled test tube and put them in the test tube rack. Put the dropper back into the beaker from which the feed sample was taken.

6. Put ten drops of distilled water into the test tube labeled as water. It will serve as a control.

7. Hypothesize which feed contains which nutrients and record your answer in Table 15.1.1 in the Results and Discussion section.

8. Record the color of each test tube of feed in Table 15.1.2.

9. Add three drops of iodine to each of the test tubes. **Caution - Use great care in handling iodine. Wash it from your skin and call your teacher immediately if any is spilled.**

10. Gently swirl each test tube to mix the contents. If the color changes to a deep blue or a blue-black, starch is present. Record the results for each sample in Table 15.1.2.

11. Follow your teacher's directions in disposing of the contents. Wash the test tubes thoroughly.

Part II Test for Sugar

12. Label the six test tubes again if their labels were removed during cleaning.

13. Using the dropper in each of the beakers of feed, place ten drops of feed into the appropriately labeled test tube and place them in the test tube rack. Add ten drops of distilled water to the control test tube. Record the color of each feedstuff in Table 15.1.3.

14. Put twenty drops of Benedict's solution into each test tube. **Caution - Use great care in handling Benedict's solution. Wash it from your skin and call your teacher immediately if any is spilled.**

15. Gently swirl each test tube to mix the solutions. Record the color in Table 15.1.3.

16. Using a test tube holder, place the test tubes in a hot water bath, as shown in Figure 15.1.2. Allow them to sit in the boiling water for five minutes.

17. Use a test tube holder to remove the six samples from the hot water bath. Place them in a test tube rack. Turn the hot plate off.

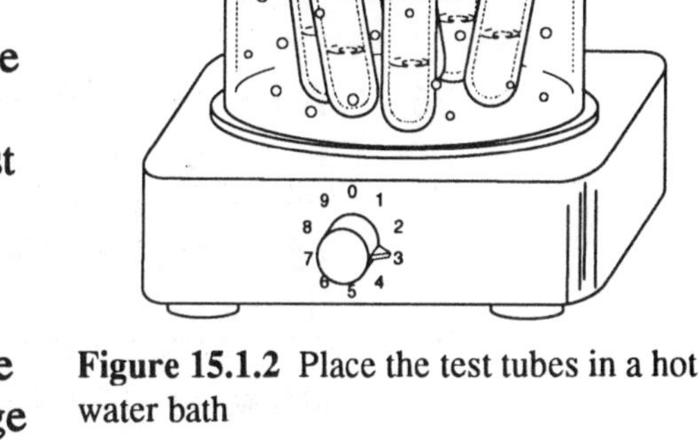

Figure 15.1.2 Place the test tubes in a hot water bath

18. Record the color of each sample in Table 15.1.3. A color change to orange, red, yellow, or green indicates the presence of sugar. Record whether or not sugar is present in each sample.

19. Follow your teacher's directions in disposing of the contents. Wash the test tubes thoroughly.

Part III Test for Protein

20. Repeat steps 12 and 13 to prepare the test tubes of feed materials for protein analysis. Record their color in Table 15.1.4.

21. Add ten drops of biuret (BY-uh-ret) solution to each test tube and swirl gently to mix the contents. **Caution - Use great care in handling biuret solution. Wash it from your skin and call your teacher immediately if any is spilled.**

22. Record the color of each test tube in Table 15.1.4 in the Results and Discussion section. The presence of protein is indicated by a shade of purple.

23. Dispose of the contents of the test tubes as directed by your teacher. Wash the test tubes thoroughly.

Part IV Test for Fat

24. With your pencil draw lines dividing a piece of brown paper into six sections. Label the sections 1 through 6. Write in the corresponding feed material. See Figure 15.1.3.

25. Use the dropper in each beaker to place a drop of feed solution in the center of the appropriate section of the paper. Place a drop of distilled water in the center of section 6.

26. After the solutions have soaked into the paper (about five minutes), wipe away any solid particles and wave the paper in the air to assist in drying.

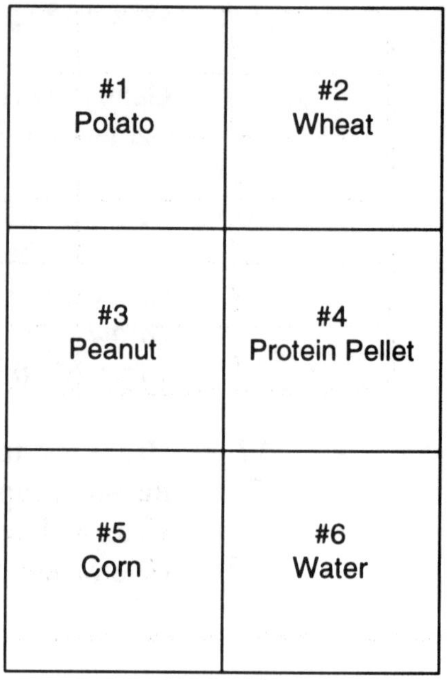

#1 Potato	#2 Wheat
#3 Peanut	#4 Protein Pellet
#5 Corn	#6 Water

Figure 15.1.3 Place a sample of each food on brown paper

27. When the spots have dried, hold the paper up to a window or bright light. If the spot is translucent (light passes through it), then the feed contains fat. Record your answers in Table 15.1.5.

28. Place the paper in the trash can and dispose of the beakers of feedstuff as directed by your teacher. Wash all equipment thoroughly.

RESULTS AND DISCUSSION

1. Record your hypothesis of what nutrients you think each feedstuff contains in Table 15.1.1. Select from starch, sugar, protein, and fat.

Table 15.1.1 Hypothesis		
Test Tube Number	Feed Material	Nutrients You Expect to Find
1		
2		
3		
4		
5		
6	Control (water)	

2. Record your findings from Part I of this laboratory exercise in Table 15.1.2. Record any unusual conditions, problems, etc., in the "Observations" column.

Table 15.1.2 Test for Starch					
Test Tube	Feedstuff	Color before Adding Iodine	Color after Adding Iodine	Is Starch Present?	Observations
1					
2					
3					
4					
5					
6	Control (water)				

3. Indicate your findings for the test for sugar in Table 15.1.3. Record any unusual conditions, problems, etc., in the "Observations" column.

Table 15.1.3 Test for Sugar						
Test Tube	Feedstuff	Color before Treatment	Color after heating with Benedict's Solution	Color after Heating with Benedict's Solution	Is Sugar Present?	Observations
1						
2						
3						
4						
5						
6	Control (water)					

4. Indicate your findings for the test for protein in Table 15.1.4. Record any unusual conditions, problems, etc., in the "Observations" column.

Table 15.1.4 Test for Protein					
Test Tube	Feedstuff	Color before Treatment	Color after Adding Biuret Solution	Is Protein Present?	Observations
1					
2					
3					
4					
5					
6	Control (water)				

5. Record the results of your test for fat in the feedstuff in Table 15.1.5. Record any unusual conditions, problems, etc. in the "Observations" column.

Table 15.1.5 Test for Fat

Test Tube	Feedstuff	Was the Spot Translucent After Drying?	Is Fat Present in the Feed?	Observations
1				
2				
3				
4				
5				
6	Control (water)			

6. What was the purpose of the test tube with distilled water?

7. What conclusion is usually drawn when there is a color change after adding an indicator solution?

NAME:________________________ DATE:______________ CLASS:____________

Chapter 16: Meat Science

Laboratory Exercise 16.1 Quality Grading Beef

BACKGROUND

As a representative sample of one phase of the meat industry, you will learn the basics of quality grading beef. This is one of the jobs performed by USDA meat graders. Federal inspectors in processing plants put each carcass into one of the classes for quality. Quality in this case is an estimate of the beef's tenderness, juiciness, and flavor. Quality grading should not be confused with yield grading, which is an estimate of edible meat in relation to carcass weight.

There are seven USDA grades for beef quality. The four most common and highest quality grades are Prime, Choice, Select, and Standard. USDA inspectors consider five factors in determining quality grade. They are maturity, marbling, color of lean, texture, and firmness of lean. The meat of cattle fed for slaughter usually have the proper color, texture, and firmness. That means that the quality grade of most meat is based on maturity and marbling.

Older animals usually have tougher meat than younger animals. Meat graders determine the age of an animal by inspecting a cross section of the spinal column. The degree of ossification of the bones gives an estimate of the animal's age. A lot of cartilage indicates a younger animal. Cartilage that has ossified (turned to bone) indicates older animals. Animals between nine and thirty months of age are eligible for the higher quality grades - Prime, Choice, Select, or Standard. Most cattle fed for slaughter fall into this age category (see Figure 16.1.1).

> Bone Ossification in Humans
>
> The amount of ossification in the bones of animals has many uses. Many high school students currently have braces on their teeth or have just had them removed. Many will remember an early visit to the orthodontist in preparation for braces when an X-ray was made of their hand and wrist. An orthodontist can estimate the maturity of a human body - and how much more growing the teeth will do - by the amount of cartilage in the hand and wrist. In a similar manner, forensic detectives can determine the age of a murder victim, even after the body has been decomposed, by examining the skeletal remains for ossification of the bones.

Marbling is the primary factor used to determine quality grade of fed cattle. Marbling is the intermingling of fat with the muscle tissue. Marbling contributes to the eating satisfaction of beef. It gives the perception of tenderness when chewed.

There are nine degrees of marbling used by the USDA, ranging from "practically devoid" to "abundant." An estimate of marbling is made by examining a rib eye cross section from the carcass. You will have an opportunity to estimate the degree of marbling in rib eyes during this exercise.

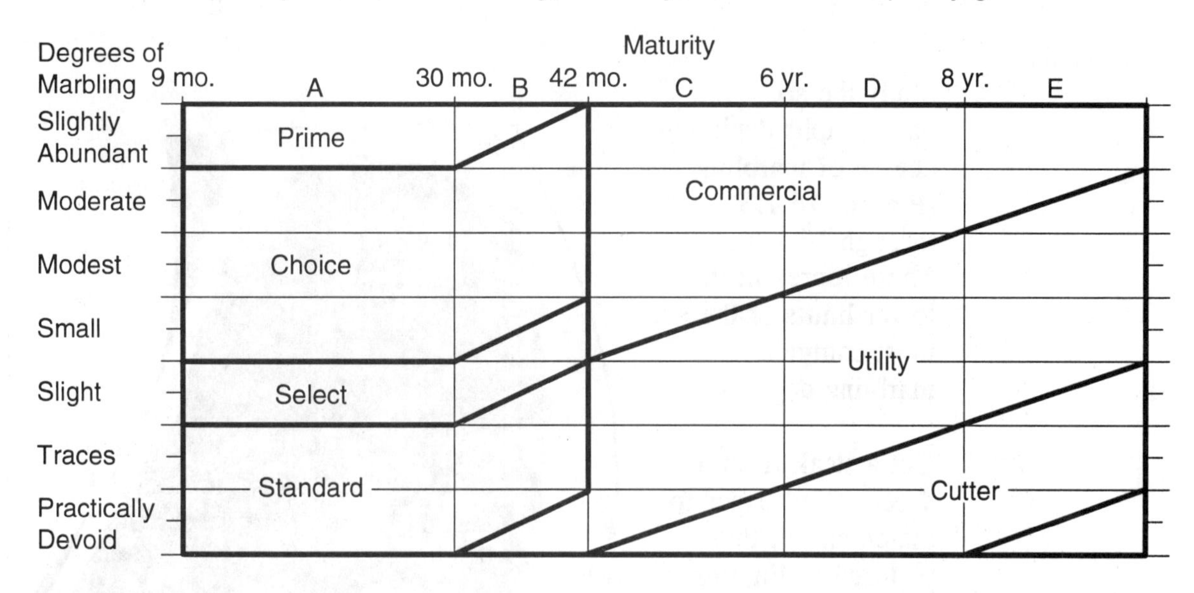

Figure 16.1.1 Courtesy of USDA

OBJECTIVES

- To identify the quality grades of beef
- To estimate the marbling of beef
- To grade beef for quality

EQUIPMENT

scissors
ruler
small craft paint brush

MATERIALS

one sheet of red paper
one sheet of black paper
white craft paint

SAFETY

There are no special safety concerns involved in this laboratory exercise. You may want to wear gloves and a laboratory apron while using paint.

PROCEDURES

1. Study the six photographs depicting degree of marbling (Figures 16.1.3 through 16.1.8). These represent the lower limits of the six most common marbling degrees.

2. Cut a steak from a sheet of red paper to match any of those pictured in Figures 16.1.3 through 16.1.8. Use a rule to help you get very close to the original size. You may want to trace the rib eye onto a sheet of paper first.

3. Paste up your rib eye onto another color of paper. Use any color that gives good contrast. Black is suggested.

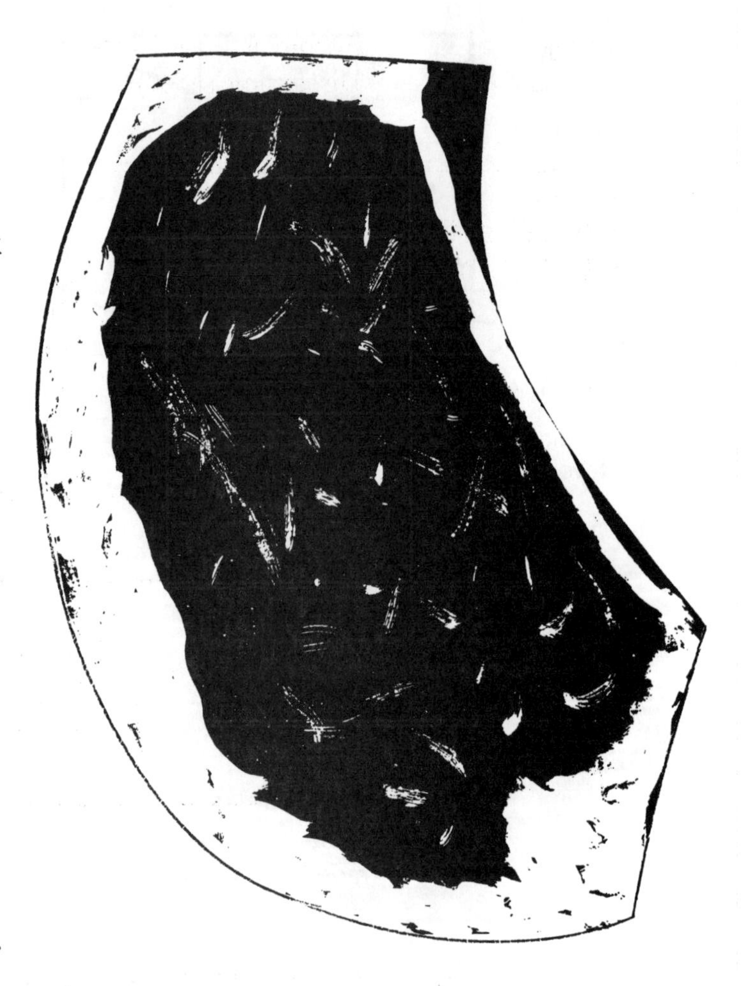

Figure 16.1.2 Your home-made steak should look similar to the one in this drawing

4. With a small craft paint brush, outline the fat covering on each rib eye with white paint. Use the photograph of the steak you are trying to copy as a guide. Make your rib eye as realistic as possible.

5. Use a small paint brush and white paint to add the marbling.

6. Write the number your teacher assigns to your rib eye at the top of the page.

7. When instructed to do so by the teacher, lay your rib eye out at your work station for viewing by your classmates. Each of your classmates will do the same.

8. The photographs in Figures 16.1.3 through 16.1.8 are standards for marbling established by the USDA. Use these photographs to estimate the degree of marbling in the rib eyes prepared by your classmates.

9. Record your answers in Table 16.1.1 provided in the Results and Discussion section.

RESULTS AND DISCUSSION

1. Is grading for quality an exact measure? Explain your answer.

2. List the five factors of concern in quality grading beef.

   1. ______________________________
   2. ______________________________
   3. ______________________________
   4. ______________________________
   5. ______________________________

3. In grading beef from animals between the ages of nine and thirty months fed for slaughter, the degree of marbling is generally the only concern for quality grade. Why is this true?

4. What is marbling?

5. How can the age of an animal be determined by its bones?

6. What should you look for the next time you are selecting a steak from the supermarket meat counter?

7. Use the work sheet below to record the degree of marbling and the quality grade you assign to each sample presented. Assume that this beef is from cattle less than thirty months old and that the texture, firmness, and color are acceptable for the highest quality grades. Use Table 16.1.1 to determine grade, based on degree of marbling.

Table 16.1.1 Marbling and Quality Grade Rating Sheet

Sample Number	Degree of Marbling	Quality Grade
1		
2		
3		
4		
5		
6		
7		
8		
9		
10		
11		
12		
13		

Sample Number	Degree of Marbling	Quality Grade
14		
15		
16		
17		
18		
19		
20		
21		
22		
23		
24		
25		
26		

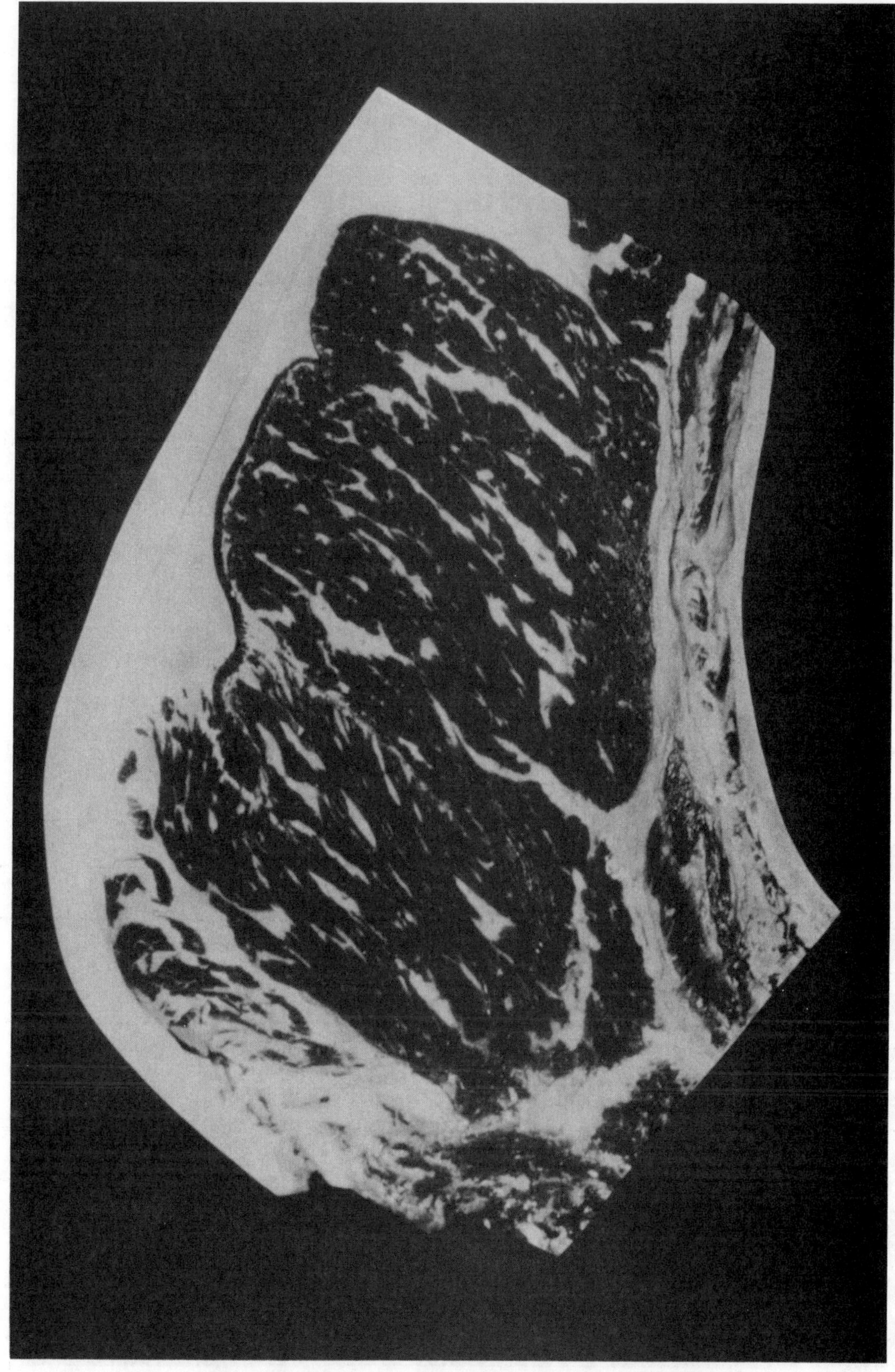

Moderately Abundant (MdA^{0})

Figure 16.1.3 Moderately Abundant (photo courtesy of National Live Stock and Meat Board)

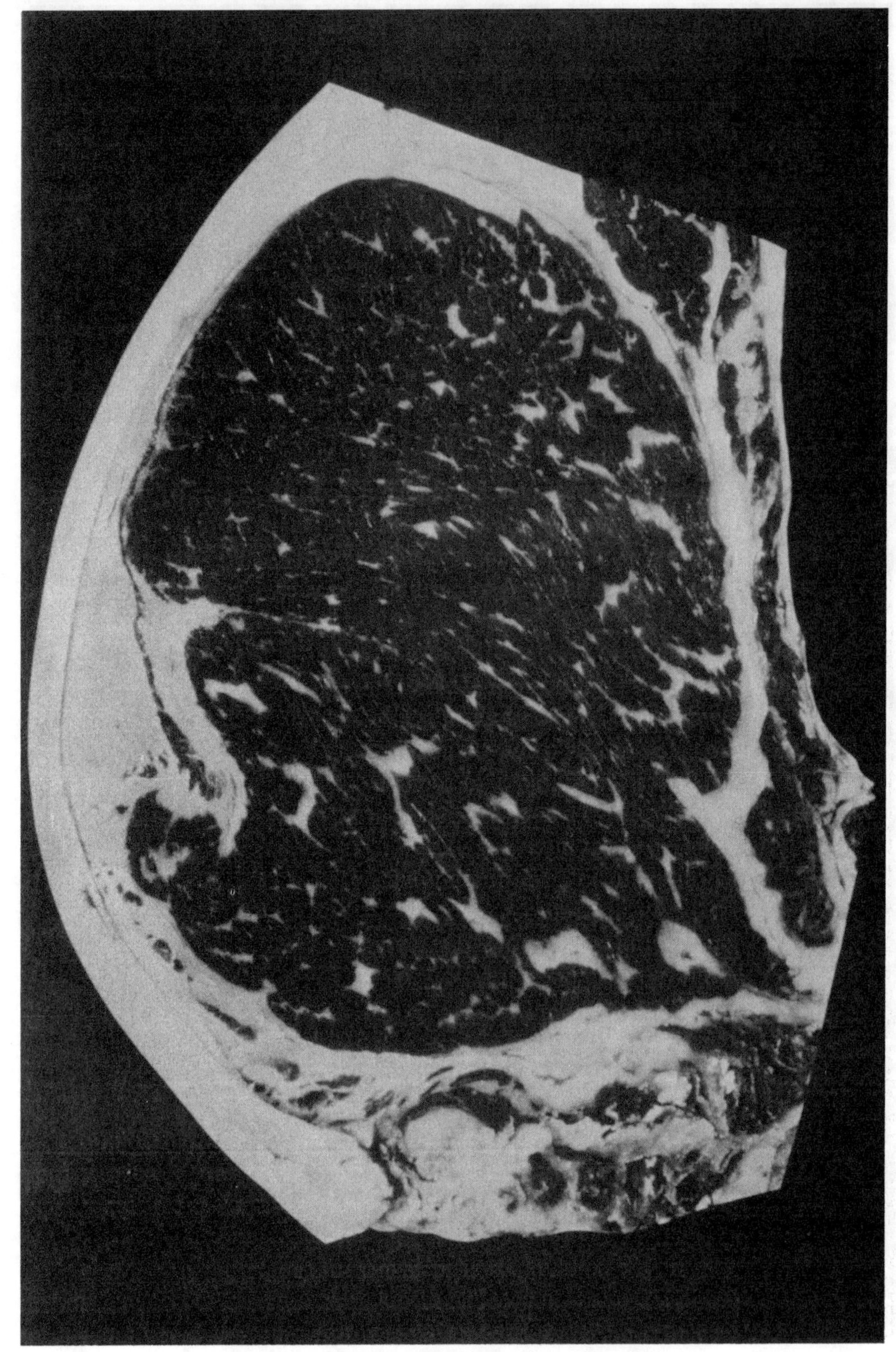

Slightly Abundant (SlA^{o})

Figure 16.1.4 Slightly Abundant (photo courtesy of National Live Stock and Meat Board)

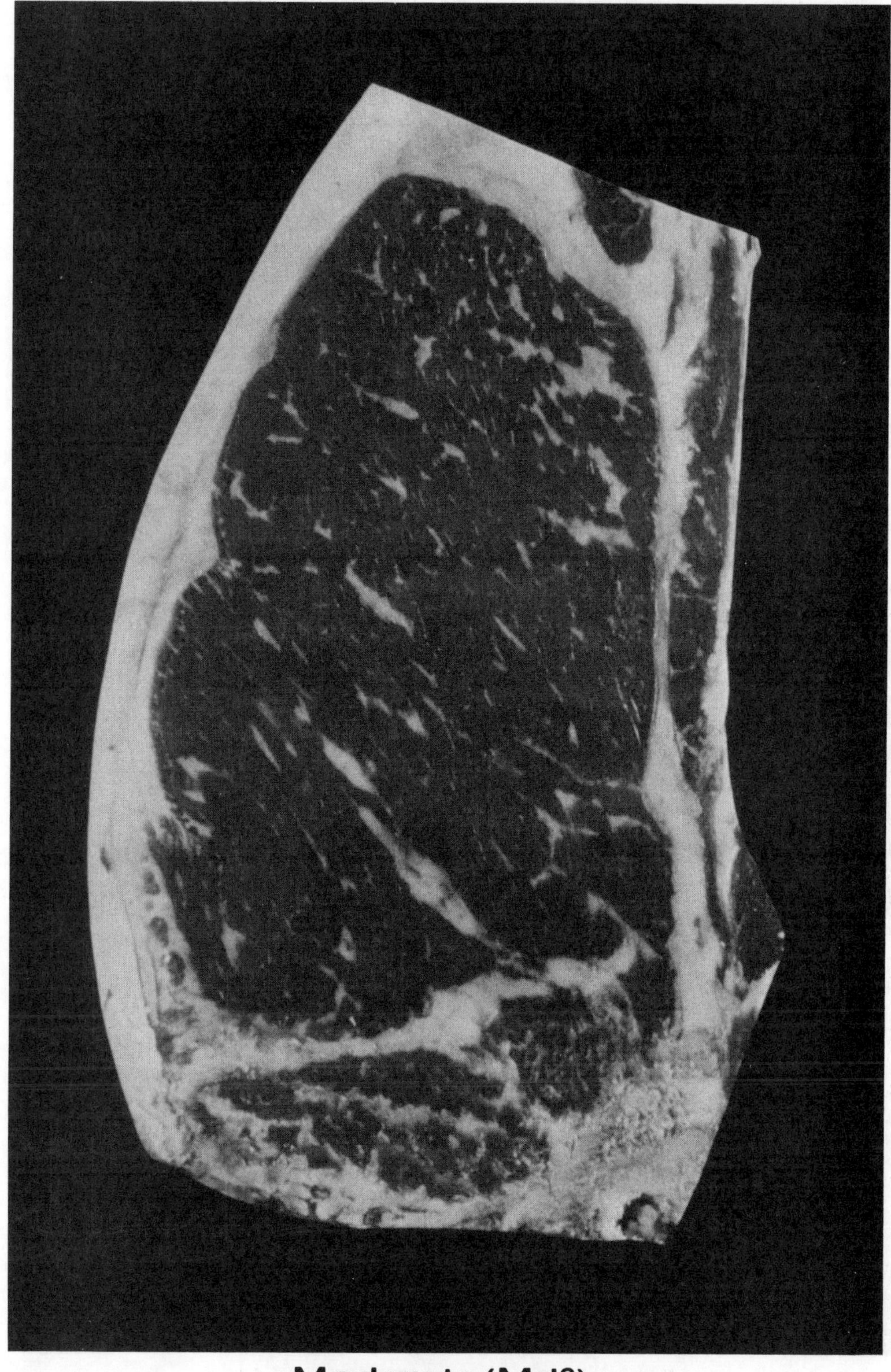

Figure 16.1.5 Moderate (photo courtesy of National Live Stock and Meat Board)

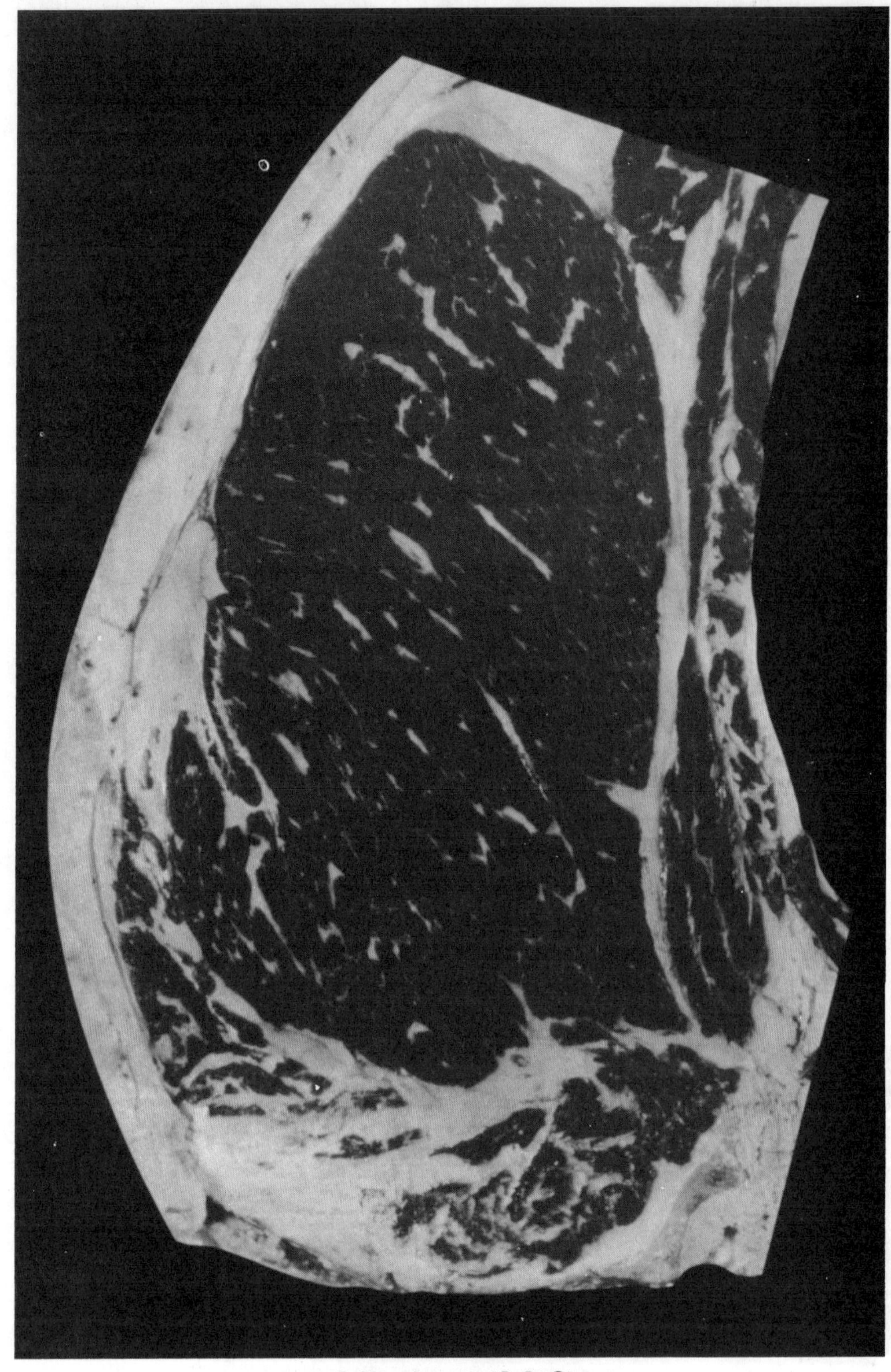

Modest(Mt°)

Figure 16.1.6 Modest (photo courtesy of National Live Stock and Meat Board)

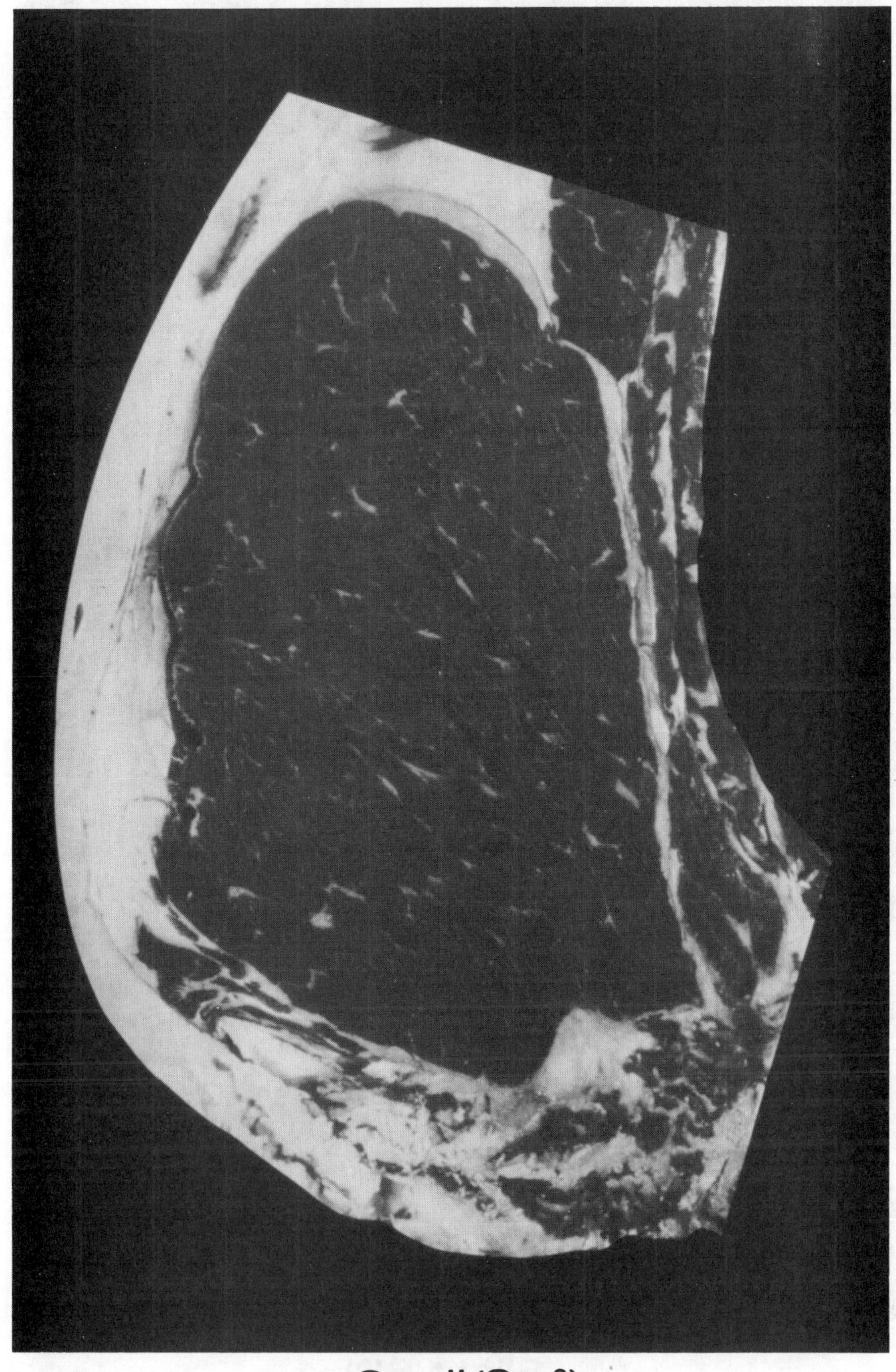

Small (Sm^{0})

Figure 16.1.7 Small (photo courtesy of National Live Stock and Meat Board)

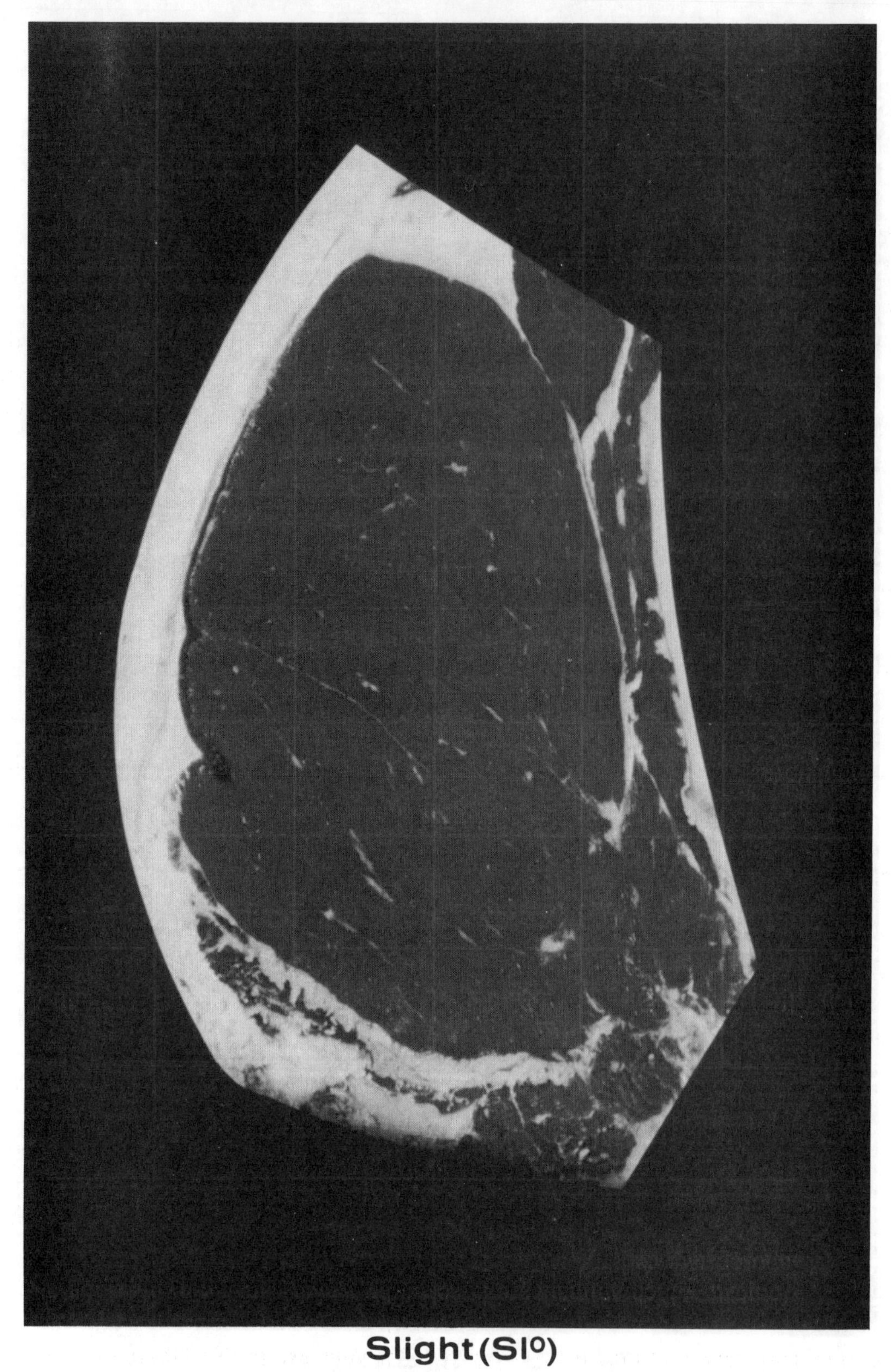

Figure 16.1.8 Slight (photo courtesy of National Live Stock and Meat Board)

NAME:________________________ DATE:____________ CLASS:____________

Chapter 16: Meat Science

Laboratory Exercise 16.2 Grading Beef for Yield

BACKGROUND

Yield grading is one of the jobs performed by USDA meat graders. Federal inspectors in processing plants put each carcass into one of the classes for yield. Yield grades are an estimate of edible meat in relation to carcass weight. The yield grade is the estimated percentage of boneless, closely trimmed, retail cuts that come from the major lean primals (round, loin, rib, chuck). USDA Yield Grades for beef are shown in Table 16.2.1.

Table 16.2.1 Yield Grades for Beef

Yield Grade	Grade Description
1	over 52.3 percent lean primal cuts
2	50.0 - 52.3 percent lean primal cuts
3	47.7 - 50.0 percent lean primal cuts
4	45.4 - 47.7 percent lean primal cuts
5	less than 45.4 percent lean primal cuts

The yield grade is determined by a formula. Factors used in the formula are the chilled carcass weight, the amount of internal (kidney, pelvic, and heart) fat, the size of the rib-eye area, and the amount of backfat on the carcass.

The yield grade formula may look complicated to the beginner. The formula calls for adjustments based on several factors as just described. In this laboratory exercise, the steps are broken down into small, manageable parts. You will determine the yield grade for six beef carcasses.

Optional Activities

Your teacher may arrange to have fresh rib eyes for your use in this laboratory. Frozen steaks kept in the school freezer work well. Most stores sell meat that is trimmed to 1/4" or less of fat. If steaks are to be purchased, be sure to ask the butcher to cut untrimmed steaks. Some classes cut "steaks" from wood on the band saw. Students outline the parts of steaks onto the wood by looking at real steaks as a guide. The wooden steaks are then painted. They are used in this laboratory exercise and others concerning meat science.

OBJECTIVES

- To define yield grade for beef
- To explain the value of high versus low yield grades
- To calculate yield grades for beef
- To estimate adjustments to yield grades for beef

EQUIPMENT

ruler marked in tenths of an inch
transparency of a rib-eye measurement grid
calculator (optional)

MATERIALS

one sheet of transparency film

SAFETY

There are no special safety precautions involved in this laboratory exercise. However, if the teacher chooses to have the students create wooden steaks as suggested in the teaching tip, students should follow carefully the instructions given by the teacher on use of equipment and should work with equipment only under the supervision of the teacher.

PROCEDURES

1. Determine the preliminary yield grade (PYG) of each of the rib eyes pictured in Figures 16.1.3 through 16.1.8 by measuring the fat thickness at a point three-quarters the length of the eye. See Figure 16.2.1. Use a ruler divided into tenths of an inch to make this measurement. Record the fat thickness for each sample in Table

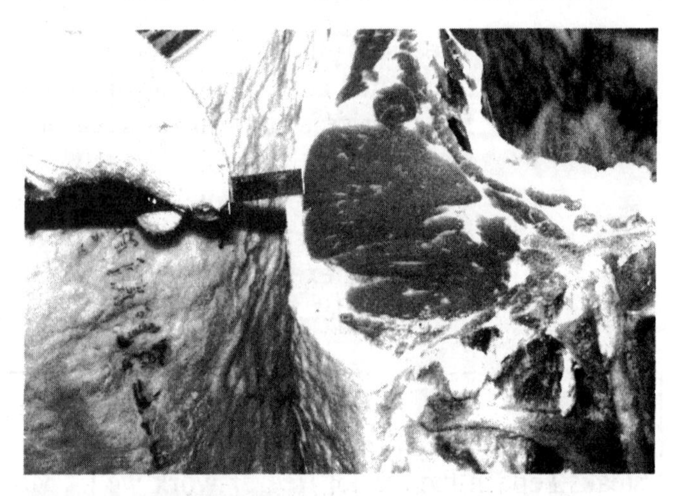

Figure 16.2.1 Measure fat thickness at a point 3/4 the length of the eye

16.2.5 of item 1 in the Results and Discussion section. Read the Preliminary Yield Grade from Table 16.2.4 and record your answers in Table 16.2.5.

2. Measure the eye of each steak in Figures 16.1.3 through 16.1.8. Obtain a transparency of the rib-eye grid measuring device from your teacher. The teacher may direct you to make a transparency grid by using the master in Figure 16.2.3, a blank piece of transparency film, and a copy machine.

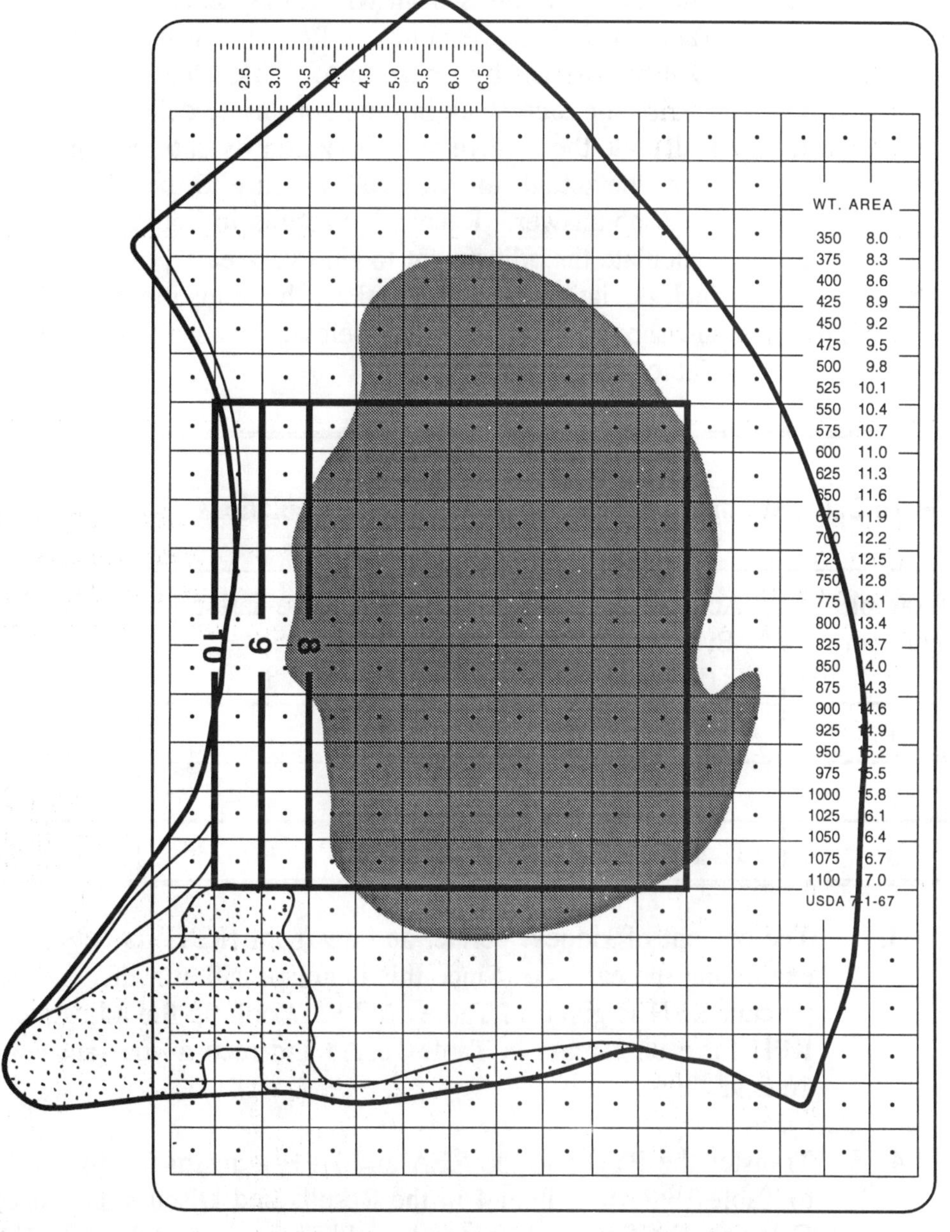

Figure 16.2.2 Rib eye area is determined by laying a grid over the eye, counting dots within the eye and dividing by 10. This rib eye measures 10.4 inches. (not drawn to scale)

a. Lay the transparency grid over each steak. See Figure 16.2.2.
b. Count the squares included in the eye of the steak. Count a square only if the center dot is within the rib-eye muscle. Muscle around the edges of the steak should not be counted.
c. Divide the number of squares by ten to determine the rib-eye area in square inches. On the grid, ten squares equal one square inch.
d. Record the rib-eye area for each steak in Table 16.2.7 in item 2 in the Results and Discussion section.
e. The carcass weights from which the steaks were cut are given in Table 16.2.7. Determine the Required Rib-Eye Area (RREA) for each carcass by referring to Table 16.2.6. Record the RREA for each steak in the appropriate cell in Table 16.2.7.
f. Calculate the rib-eye area deviation by subtracting the RREA from the actual rib-eye area. Be sure to include a + or - sign for each answer. Record the results in Table 16.2.7.
g. Calculate the adjustment to rib-eye area. For every 0.3-square-inch deviation from the RREA, the adjustment is 0.1. Examples are shown in Table 16.2.2 below. Record your answers in Table 16.2.7.

Table 16.2.2 Examples of Rib-Eye Adjustment Calculations

Example Number	Actual Rib Eye Area (REA)	Required Rib Eye Area	Deviation	Adjustment
1	11.2	11.6	-0.4	+0.1
2	10.9	12.3	-1.4	+0.5
3	12.4	11.8	+0.6	-0.2

3. The percent of kidney, pelvic, and heart fat (KPH) is determined by examining the carcass. Since this is not possible in this exercise, the percent KPH is given in Table 16.2.9. Look up the adjustment for KPH for each carcass in Table 16.2.8 and record your answer in Table 16.2.9.

4. Transfer the PYG and the REA and KPH adjustments for each rib eye to Table 16.2.10 in item 4 in the Results and Discussion section. Calculate the final yield grade by adding across. Add the REA and KPH adjustments to the PYG. Record your answers in the Final Yield Grade column of Table 16.2.10. Two examples are shown in Table 16.2.3 to help you in this step.

Table 16.2.3 Examples of Determining Final Yield Grade				
Example Number	PYG	REA Adjustment	KPH Adjustment	Final Yield Grade
1	2.5	-.7	-.2	1.6
2	3.0	+.2	-.3	2.9

RESULTS AND DISCUSSION

1. Complete Table 16.2.5 below on preliminary yield grade (PYG) for each of the six sample rib eyes pictured in Figures 16.1.3 through 16.1.8. Read the PYG from Table 16.2.4.

Table 16.2.4 Preliminary Yield Grade (PYG)	
Fat Thickness in Inches	PYG
0.0	2.00
0.1	2.25
0.2	2.50
0.3	2.75
0.4	3.00
0.5	3.25
0.6	3.50
0.7	3.75
0.8	4.00
0.9	4.25
1.0	4.50

Table 16.2.5 Preliminary Yield Grade Calculation		
Sample Number	Fat in Inches	Preliminary Yield Grade (PYG)
1		
2		
3		
4		
5		
6		

2. Complete Table 16.2.7 on rib-eye adjustment for the six sample rib eyes. Read the required rib-eye area from Table 16.2.6.

Table 16.2.6 Rib-Eye Area Adjustment Table	
Carcass Weight	Required Rib Eye Area
500	9.8
525	10.1
550	10.4
575	10.7
600	11.0
625	11.3
650	11.6
675	11.9
700	12.2

Table 16.2.7 Rib-Eye Adjustment					
Sample	Carcass Weight	Required Rib-Eye Area (RREA)	Actual Rib-Eye Area (REA)	Rib-Eye Area Deviation (+,-)	Rib-Eye Area (REA) Adjustment (+,-)
1	675				
2	625				
3	550				
4	675				
5	650				
6	700				

3. The percent of kidney, pelvic, and heart fat (KPH) for each sample rib eye is given in Table 16.2.9. Indicate the adjustment that must be made for each rib eye by looking up the % KPH in Table 16.2.8.

Table 16.2.8 KPH Adjustment Table	
KPH	Adjustment
0.5	-0.6
1.0	-0.5
1.5	-0.4
2.0	-0.3
2.5	-0.2
3.0	-0.1
3.5	0.0
4.0	+0.1
4.5	+0.2
5.0	+0.3
5.5	+0.4

Table 16.2.9 Calculating KPH Adjustment		
Sample Number	% Kidney, Pelvic, and Heart Fat (KPH)	Kidney, Pelvic, and Heart Fat Adjustment (KPH)
1	3.5	
2	3.0	
3	2.5	
4	2.0	
5	1.5	
6	4.5	

4. Complete Table 16.2.10. Determine the Final Yield Grade (FYG) by entering the adjustments determined in items 1 through 3. Add across the table to calculate the FYG for each rib eye.

Table 16.2.10 Determining Final Yield Grade				
Sample Number	PYG	REA Adjustment	KPH Adjustment	Final Yield Grade (FYG)
1				
2				
3				
4				
5				
6				

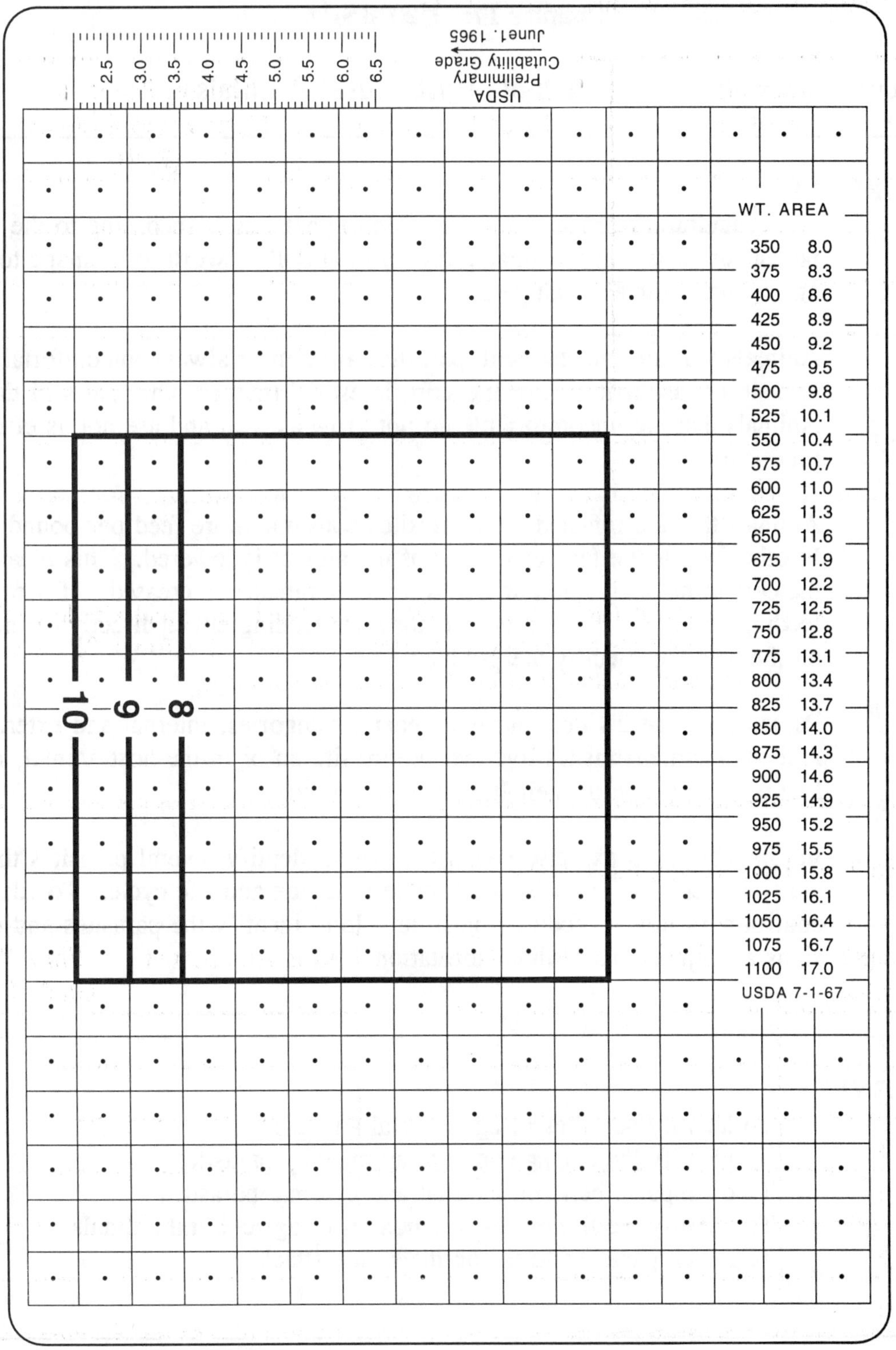

USDA
Preliminary
Cutability Grade
June 1, 1965

2.5 3.0 3.5 4.0 4.5 5.0 5.5 6.0 6.5

10 9 8

WT.	AREA
350	8.0
375	8.3
400	8.6
425	8.9
450	9.2
475	9.5
500	9.8
525	10.1
550	10.4
575	10.7
600	11.0
625	11.3
650	11.6
675	11.9
700	12.2
725	12.5
750	12.8
775	13.1
800	13.4
825	13.7
850	14.0
875	14.3
900	14.6
925	14.9
950	15.2
975	15.5
1000	15.8
1025	16.1
1050	16.4
1075	16.7
1100	17.0

USDA 7-1-67

Figure 16.2.3 Rib eye grid transparency master

NAME:______________________ DATE:____________ CLASS:____________

Chapter 17: Parasites

Laboratory Exercise 17.1 Parasites of Agricultural Animals

BACKGROUND

All agricultural animals are susceptible to parasites. According to the USDA, parasitism causes approximately one billion dollars worth of damage to agricultural animals each year.

Animals that are infected with parasites are almost always uncomfortable. Parasites cause irritation of the skin, intestinal tract or other parts of the body. Animals that are uncomfortable do not grow as well and are not as efficient in production.

Animals that are infected with parasites consume more feed per pound of gain. In other words the feed efficiency of the animal is lowered. This means that the cost to maintain and put weight on the animal is increased. If parasites are feeding on an agricultural animal, they are feeding either directly or indirectly on the feed supplied by the producer.

Parasites can be divided into two general categories: internal and external. Although some parasites live their entire life on or in the host animal, most live only a portion of their life on or in the host.

In this laboratory exercise you will learn to identify several parasites that attack agricultural animals, their mode of attack and life cycle. To adequately control parasites, a producer must be able to identify the parasites and the signs and symptoms of their infestation.

OBJECTIVES

- To identify parasites of agricultural animals
- To identify the damage done to animals by parasites
- To distinguish between internal and external parasites
- To identify the life cycles of parasites of agricultural animals
- To match parasites to the animals they attack

EQUIPMENT

dissection microscope

MATERIALS

Text - *The Science of Animal Agriculture*
reference materials on parasites (optional)
preserved parasite specimens

- roundworm
- heel fly
- biting fly
- louse
- tick
- mite
- fluke
- strongyles
- tapeworm

petri dish or slide for each parasite

SAFETY

There are no inherent dangers associated with this laboratory exercise, although general laboratory safety procedures should be followed.

PROCEDURES

1. Obtain a preserved specimen of one of the parasites selected by your teacher for the laboratory. The teacher may select any or all of the pests listed in the materials section.

2. Observe each parasite either under a dissection microscope or with a hand lens. Sketch each parasite in the space provided.

3. Review the text and teacher-supplied reference materials to complete the matrix on parasites in the Results and Discussion section.

4. Return your parasite to the original container.

5. Repeat steps 1 through 4 until you have completed the exercise for each of the parasites the teacher has selected.

RESULTS AND DISCUSSION

1. Sketch each parasite as observed under a dissection microscope or with a hand lens.

1. ____________ Parasite name	2. ____________ Parasite name	3. ____________ Parasite name
4. ____________ Parasite name	5. ____________ Parasite name	6. ____________ Parasite name
7. ____________ Parasite name	8. ____________ Parasite name	9. ____________ Parasite name

2. Indicate which animals are attacked by which parasite by marking an *X* in the appropriate cell.

Table 17.1.1 Parasites of Agricultural Animals

Parasite	Swine	Cattle	Sheep	Horses	Poultry
Tick					
Roundworm					
Louse					
Mite					
Heel fly					
Biting Fly					
Fluke					
Strongyle					
Tapeworm					

3. Complete the matrix for each parasite observed in this laboratory. Put an *X* in each block that applies to the parasite named in the left column.

Table 17.1.2 Characteristics of Parasites of Agricultural Animals

Parasite	Is an internal parasite	Is an external parasite	Adult stage is a worm	Adult stage is a fly	Adult stage is a beetle	Sucks blood	Eats feed in intestine	Completes entire life cycle on the animal	Completes part of life cycle off host animal	Immature stage is a larva	Has six legs (insect)	Immature stage is a nymph
Tick												
Roundworm												
Louse												
Mite												
Heel fly												
Biting Fly												
Fluke												
Strongyle												
Tapeworm												

4. Why is it important to control parasites of agricultural animals? Explain how parasites cause economic loss in animal production.

5. Explain how the life cycle of a parasite may be used to control a parasite. (You may need to refer to chapter 17 of the text to answer this question.)

NAME:____________________________ DATE:______________ CLASS:_____________

Chapter 18: Diseases

Laboratory Exercise 18.1 Culturing Bacteria

BACKGROUND

Bacteria are microscopic one-celled organisms. While many bacteria are beneficial to man, some bacteria spoil food, and others cause diseases of plants and animals. Individual bacteria are so small that they must be magnified many times to be seen with the human eye. Bacteria are constantly present in the environment, most of the time without you being aware of them.

The purpose of this experiment is to help you understand the sources of bacteria. Even though the bacteria you culture are not necessarily those that cause animal diseases, these organisms are very common in the environment. You will become aware of the frequent exposure of animals to disease-causing organisms. In this experiment you will culture bacteria so that you may see the growth of a bacteria colony. There are thousands of bacteria in a bacteria culture large enough to be seen with the naked eye.

OBJECTIVES

- To determine sources of bacteria in the environment
- To observe a living culture of bacteria

EQUIPMENT

one sterile petri dish with nutrient agar
one beaker or glass for alcohol
grease marking pencil or felt tip pen

MATERIALS

three sterile cotton swabs
alcohol

SAFETY

- Do not open the petri dishes after the experiment has begun.
- Wash your hands after completing this exercise.
- You should never eat or drink in the laboratory.

PROCEDURES

1. Obtain a sterile petri dish with nutrient agar from your teacher. Keep the dishes closed until directed to open them.

2. Turn the dish over on the table and divide the bottom into fourths by drawing on the bottom with a grease pencil, as shown in Figure 18.1.1. Number each section. Put your initials on one section of the dish for easy identification.

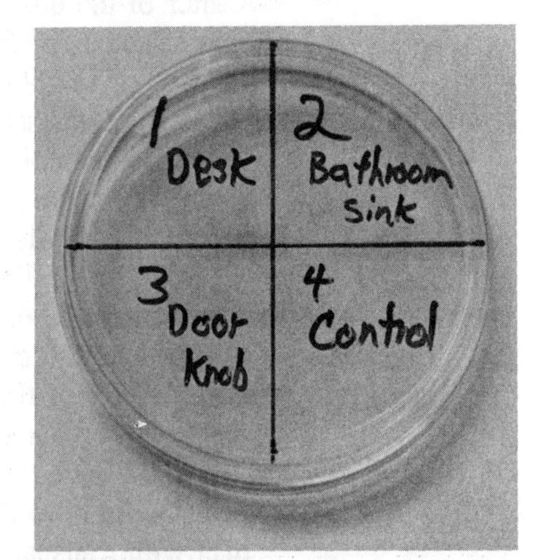

Figure 18.1.1 Divide the petri dish into four sections

3. You are now ready to collect three sources of bacteria from the environment. The fourth quadrant of the dish will not be inoculated. It will serve as a control. Select one source that appears to be clean, such as a lunchroom tabletop. The other sources may be areas such as your hand, a doorknob, a book, the rim of a fish tank, etc. You may wish to use a piece of your hair.

4. Record your sources in the Results and Discussion section **before** you collect the samples.

5. Obtain a clean cotton swab. A newly opened package should be used. Take great care to keep one end of the swab sterile. Do not touch the sterile end or allow it to come into contact with anything except your source of bacteria.

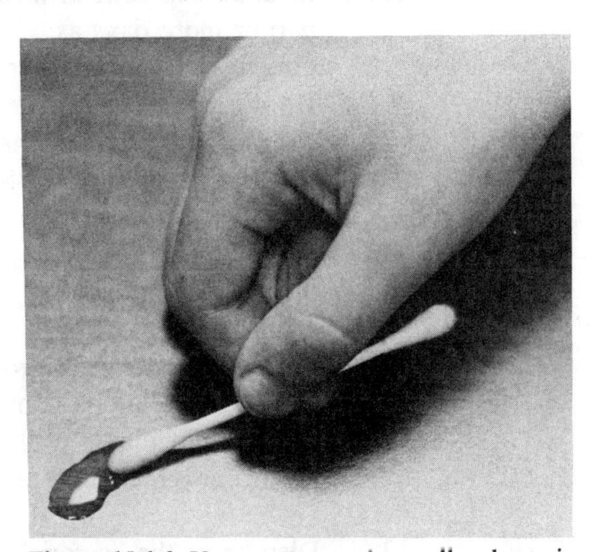

Figure 18.1.2 Use a cotton swab to collect bacteria

6. Put a drop of clean water on one of your chosen sources of bacteria. Gently rub the sterile end of the swab through the drop several times. (See Figure 18.1.2.)

7. Lift the lid of the petri dish only far enough to allow you to insert the cotton swab. Gently touch the swab to the agar surface in the center of the quadrant you selected for that source. (See Figure 18.1.3.) Dispose of the contaminated cotton swab by dropping it into a container of alcohol.

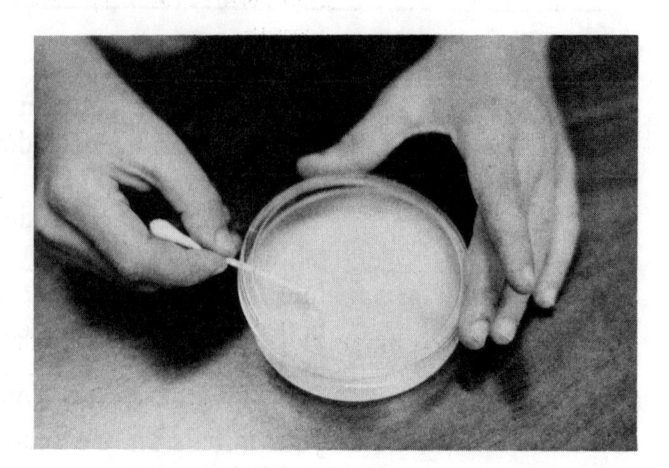

Figure 18.1.3 Lift the petri dish lid only high enough to get the cotton swab in for inoculation

8. Repeat steps 5, 6, and 7 to inoculate the other two sections of the dish. Do not inoculate section 4. This will be your control.

9. Seal the lid onto the petri dish with transparent tape. The dish should not be opened again - especially after the incubation period.

10. Incubate the petri dish for two or more days as directed by your teacher. **Do not open the petri dish after or during the incubation period.**

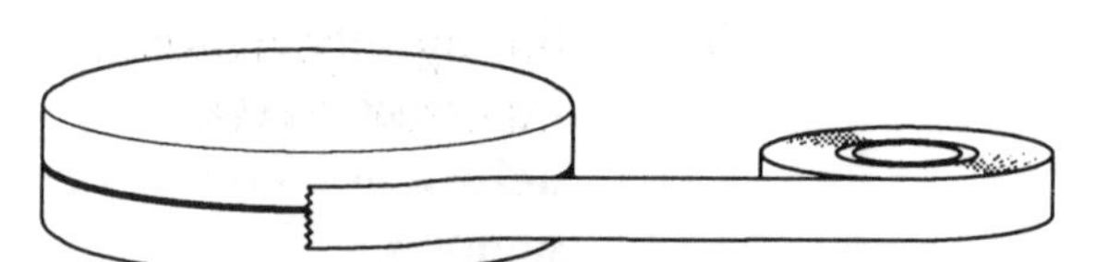

Figure 18.1.4 Seal the petri dish with tape

11. At the end of the incubation period, examine the dish for evidence of bacterial growth and record the results in the Results and Discussion section. Your colony is likely to be really icky if you got the dish inoculated well, but if not, you may have to hold the dish toward a light source to see the bacterial growth.

12. Dispose of the bacterial colony and petri dish as directed by your teacher.

RESULTS AND DISCUSSION

1. List the sources of bacteria for your colony in Table 18.1.1. List your expected results by ranking the samples from most growth to least growth.

Table 18.1.1 Bacteria Colony Sources and Expected Results

Sample Number	Source	Rank of Expected Results most growth = 1 least growth = 4
1		
2		
3		
4	Control - not inoculated	

2. Why is it necessary to tape the dishes shut and not open them once the culture has been started?

3. During inoculation of your petri dish with bacteria, why was it important to open the dish only part way?

4. Why were you instructed to label the bottoms of the petri dishes instead of the tops?

5. After incubation, draw what you see in each quadrant of the petri dish. Write a description of what you see under your drawings.

Quadrant 1

Description:

Quadrant 2

Description:

Quadrant 3

Description:

Quadrant 4 (control)

Description:

6. Was there bacteria growth in the control area? If so, where might these bacteria have come from?

7. Was your ranking of expected growth correct? Rank the actual results here and explain findings that surprised you.

Table 18.1.2 Bacteria Colony Record

Sample Number	Source	Rank of Resulting Growth most growth = 1 least growth = 4
1		
2		
3		
4	Control - not inoculated	

NOTES

NOTES

NOTES

NOTES

NOTES

NOTES

NOTES

NOTES